Advancing Maths for AQA

FURTHER MATHS 1

Sam Boardman, Tony Clough and David Evans

Series editors
Sam Boardman Roger Williamson Ted Graham David Pearson

heinemann.co.uk
✓ Free online support
✓ Useful weblinks
✓ 24 hour online ordering

01865 888058

Heinemann
Inspiring generations

Heinemann is an imprint of Pearson Education Limited, a company incorporated in England and Wales, having its registered office at Edinburgh Gate, Harlow, Essex, CM20 2JE. Registered company number: 872828

Heinemann is a registered trademark of Pearson Education Limited

First published 2004

12
10 9

British Library Cataloguing in Publication Data is available from the British Library on request.

ISBN: 978 0 435513 34 4

Edited by Alex Sharpe, Standard Eight Limited
Typeset and illustrated by Tech-Set Limited, Gateshead, Tyne & Wear.
Original illustrations © Harcourt Education Limited, 2004
Cover design by Miller, Craig and Cocking Ltd
Printed in China (CTPS/09)

Acknowledgements
The publishers' and authors' thanks are due to the AQA for permission to reproduce questions from past examination papers.

The answers have been provided by the authors and are not the responsibility of the examining board.

Every effort has been made to contact copyright holders of material reproduced in this book. Any omissions will be rectified in subsequent printings if notice is given to the publishers.

About this book

This book is one in a series of textbooks designed to provide you with exceptional preparation for AQA's 2004 Mathematics Specification. The series authors are all senior members of the examining team and have prepared the textbooks specifically to support you in studying this course.

Finding your way around

The following are there to help you find your way around when you are studying and revising:

- **edge marks** (shown on the front page) – these help you to get to the right chapter quickly;
- **contents list** – this identifies the individual sections dealing with key syllabus concepts so that you can go straight to the areas that you are looking for;
- **index** – a number in bold type indicates where to find the main entry for that topic.

Key points

Key points are not only summarised at the end of each chapter but are also boxed and highlighted within the text like this:

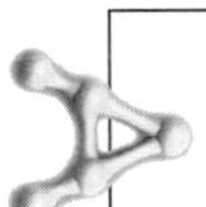

When the quadratic equation $ax^2 + bx + c = 0$ has roots α and β:

- The sum of the roots, $\alpha + \beta = -\frac{b}{a}$;
- and the product of roots, $\alpha\beta = \frac{c}{a}$.

Exercises and exam questions

Worked examples and carefully graded questions familiarise you with the syllabus and bring you up to exam standard. Each book contains:

- Worked examples and Worked exam questions to show you how to tackle typical questions; Examiner's tips will also provide guidance;
- Graded exercises, gradually increasing in difficulty up to exam-level questions, which are marked by an [A];
- Test-yourself sections for each chapter so that you can check your understanding of the key aspects of that chapter and identify any sections that you should review;
- Answers to the questions are included at the end of the book.

Contents

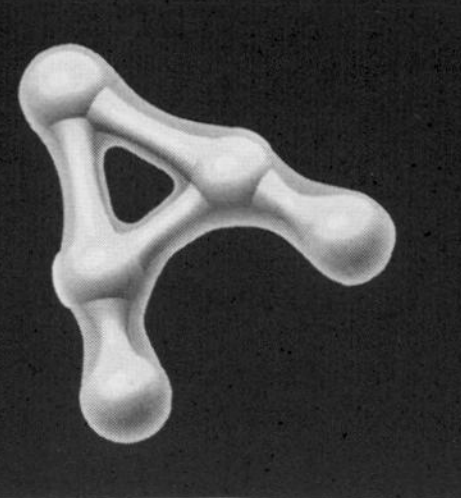

8 Calculus

9 Series

10 Numerical methods

11 Asymptotes and rational functions of the form $\frac{ax+b}{cx+d}$

CHAPTER 1

Roots of quadratic equations

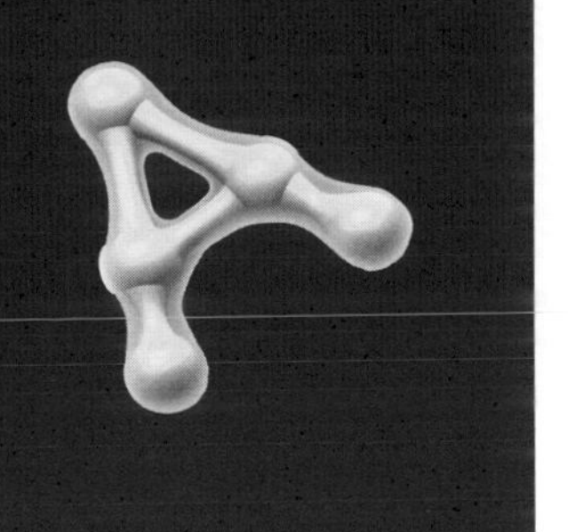

Learning objectives

After studying this chapter, you should:

- know the relationships between the sum and product of the roots of a quadratic equation and the coefficients of the equation
- be able to manipulate expressions involving $\alpha + \beta$ and $\alpha\beta$
- be able to form equations with roots related to a given quadratic equation.

In this chapter you will be looking at quadratic equations with particular emphasis on the properties of their solutions or roots.

1.1 The relationships between the roots and coefficients of a quadratic equation

As you have already seen in the C1 module, any quadratic equation will have **two roots** (even though one may be a repeated root or the roots may not even be real). In this section you will be considering some further properties of these two roots.

Suppose you know that the two solutions of a quadratic equation are $x = 2$ and $x = -5$ and you want to find a quadratic equation having 2 and -5 as its roots.

The method consists of working backwards, i.e. following the steps for solving a quadratic equation but in reverse order.

Now if $x = 2$ and $x = -5$ are the solutions then the equation could have been factorised as

$$(x - 2)(x + 5) = 0.$$

Expanding the brackets gives

$$x^2 + 3x - 10 = 0.$$

Actually, any multiple of this equation will also have the same roots, e.g.

$2x^3 + 6x - 20 = 0$

$3x^2 + 9x - 30 = 0$

$\frac{1}{2}x^2 + \frac{3}{2}x - 5 = 0$

This is a quadratic equation with roots 2 and -5.

The general case

Consider the most general quadratic equation $ax^2 + bx + c = 0$ and suppose that the two solutions are $x = \alpha$ and $x = \beta$.

Now if α and β are the roots of the equation then you can 'work backwards' to generate the original equation.

A quadratic with the two solutions $x = \alpha$ and $x = \beta$ is

$$(x - \alpha)(x - \beta) = 0.$$

Any multiple of this equation such as $kx^2 - k(\alpha + \beta)x + k\alpha\beta = 0$ will also have roots α and β.

Expanding the brackets gives

$$x^2 - \alpha x - \beta x + \alpha\beta = 0$$

$$\Rightarrow \quad x^2 - (\alpha + \beta)x + \alpha\beta = 0 \quad [1]$$

The most general quadratic equation is $ax^2 + bx + c = 0$ and this can easily be written in the same form as equation [1].

You can divide $ax^2 + bx + c = 0$ throughout by a giving

$$x^2 + \frac{b}{a}x + \frac{c}{a} = 0 \quad [2]$$

The coefficients of x^2 in [1] and in [2] are now both equal to 1.

Recall that the coefficient of x^2 is the number in front of x^2.

Since [1] and [2] have the same roots, α and β, and are in the same form, you can write

$$x^2 - (\alpha + \beta)x + \alpha\beta \equiv x^2 + \frac{b}{a}x + \frac{c}{a}$$

The $\equiv$ sign means 'identically equal to'.

Equating coefficients of x gives

$$-(\alpha + \beta) = \frac{b}{a} \quad \Rightarrow \quad \alpha + \beta = -\frac{b}{a}$$

The coefficients of x and the constant terms must be equal.

Equating constant terms gives

$$\alpha\beta = \frac{c}{a}$$

When the quadratic equation $ax^2 + bx + c = 0$ has roots α and β:

- The sum of the roots, $\alpha + \beta = -\frac{b}{a}$;
- and the product of roots, $\alpha\beta = \frac{c}{a}$.

Notice it is fairly easy to express

$x^2 + \frac{b}{a}x + \frac{c}{a} = 0$ as

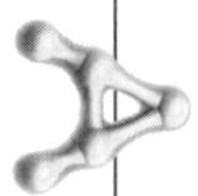

$$x^2 - (\text{sum of roots})x + (\text{product of roots}) = 0$$

Worked example 1.1

Write down the sum and the product of the roots for each of the following equations:

(a) $2x^2 + 12x - 3 = 0$, **(b)** $x^2 - 8x + 5 = 0$.

Solution

(a) $2x^2 + 12x - 3 = 0$

Here $a = 2$, $b = 12$ and $c = -3$.

Take careful note of the signs.

The sum of roots, $\alpha + \beta = -\dfrac{b}{a} = -\dfrac{12}{2} = -6$.

The product of roots, $\alpha\beta = \dfrac{c}{a} = \dfrac{-3}{2} = -1\frac{1}{2}$.

(b) $x^2 - 8x + 5 = 0$

Here $a = 1$, $b = -8$ and $c = 5$.

The sum of roots, $\alpha + \beta = -\dfrac{b}{a} = -\dfrac{-8}{1} = 8$.

Notice the double negative.

The product of roots, $\alpha\beta = \dfrac{c}{a} = \dfrac{5}{1} = 5$.

Worked example 1.2

Find the sum and the product of the roots of each of the following quadratic equations:

(a) $4x^2 + 8x = 5$, **(b)** $x(x - 4) = 6 - 2x$.

Solution

Neither equation is in the form $ax^2 + bx + c = 0$ and so the first thing to do is to get them into this standard form.

(a) $4x^2 + 8x = 5$

$\Rightarrow 4x^2 + 8x - 5 = 0$

Now in the form $ax^2 + bx + c = 0$.

In this case $a = 4$, $b = 8$ and $c = -5$.

The sum of roots, $\alpha + \beta = -\dfrac{b}{a} = -\dfrac{8}{4} = -2$.

The product of roots, $\alpha\beta = \dfrac{c}{a} = \dfrac{-5}{4} = -\dfrac{5}{4}$.

(b) $x(x - 4) = 6 - 2x$

Expand the brackets and take everything onto the LHS.

$\Rightarrow x^2 - 4x + 2x - 6 = 0$

$\Rightarrow x^2 - 2x - 6 = 0$

Now in the standard form.

Here $a = 1$, $b = -2$ and $c = -6$.

The sum of roots, $\alpha + \beta = -\dfrac{b}{a} = -\dfrac{-2}{1} = 2$.

The product of roots, $\alpha\beta = \dfrac{c}{a} = \dfrac{-6}{1} = -6$.

Worked example 1.3

Write down equations with integer coefficients for which:

(a) sum of roots = 4, product of roots = −7,

(b) sum of roots = −4, product of roots = 15,

(c) sum of roots = $-\frac{3}{5}$, product of roots = $-\frac{1}{2}$.

Solution

(a) A quadratic equation can be written as

$$x^2 - (\text{sum of roots})x + (\text{product of roots}) = 0$$

$$x^2 - (4)x + (-7) = 0$$

$$\Rightarrow \quad x^2 - 4x - 7 = 0$$

(b) Using

$$x^2 - (\text{sum of roots})x + (\text{product of roots}) = 0$$

gives $x^2 - (-4)x + (15) = 0$

$$\Rightarrow \quad x^2 + 4x + 15 = 0$$

Again great care must be taken with the signs.

(c) Using

$$x^2 - (\text{sum of roots})x + (\text{product of roots}) = 0$$

gives $x^2 - (-\frac{3}{5})x + (-\frac{1}{2}) = 0$

$$\Rightarrow \quad x^2 + \tfrac{3}{5}x - \tfrac{1}{2} = 0$$

Some of the coefficients are fractions not integers. You can eliminate the fractions by multiplying throughout by 10 or (2 × 5).

$$\Rightarrow \quad 10x^2 + 10(\tfrac{3}{5})x - 10(\tfrac{1}{2}) = 0$$

$$\Rightarrow \quad 10x^2 + 6x - 5 = 0$$

You now have integer coefficients.

EXERCISE 1A

1 Find the sum and product of the roots for each of the following quadratic equations:

(a) $x^2 + 4x - 9 = 0$ **(b)** $2x^2 - 3x - 5 = 0$

(c) $2x^2 + 10x - 3 = 0$ **(d)** $1 + 2x - 3x^2 = 0$

(e) $7x^2 + 12x = 6$ **(f)** $x(x - 2) = x + 6$

(g) $x(3 - x) = 5x - 2$ **(h)** $ax^2 - a^2x - 2a^3 = 0$

(i) $ax^2 + 8a = (1 - 2a)x$ **(j)** $\dfrac{4}{x + 5} = \dfrac{x - 3}{2}$

2 Write down a quadratic equation with:

(a) sum of roots = 5, product of roots = 8,

(b) sum of roots = −3, product of roots = 5,

(c) sum of roots = 4, product of roots = −7,

(d) sum of roots = −9, product of roots = −4,

(e) sum of roots = $\frac{1}{4}$, product of roots = $\frac{2}{5}$,

(f) sum of roots = $-\frac{2}{3}$, product of roots = 4,

(g) sum of roots = $\frac{3}{5}$, product of roots = 0,

(h) sum of roots = k, product of roots = $3k^2$,

(i) sum of roots = $k + 2$, product of roots = $6 - k^2$,

(j) sum of roots = $-(2 - a^2)$, product of roots = $(a + 7)^2$.

1.2 Manipulating expressions involving α and β

As you have seen in the last section, given a quadratic equation with roots α and β you can find the values of $\alpha + \beta$ and $\alpha\beta$ without solving the equation.

As you will see in this section, it is useful to be able to write other expressions involving α and β in terms of $\alpha + \beta$ and $\alpha\beta$.

Worked example 1.4

Given that $\alpha + \beta = 4$ and $\alpha\beta = 7$, find the values of:

(a) $\dfrac{1}{\alpha} + \dfrac{1}{\beta}$, **(b)** $\alpha^2\beta^2$.

Solution

(a) You need to write this expression in terms of $\alpha + \beta$ and $\alpha\beta$ in order to use the values given in the question.

Now $\dfrac{1}{\alpha} + \dfrac{1}{\beta} = \dfrac{\beta + \alpha}{\alpha\beta} = \dfrac{\alpha + \beta}{\alpha\beta}$

$\Rightarrow \quad \dfrac{1}{\alpha} + \dfrac{1}{\beta} = \dfrac{4}{7}$

(b) $\alpha^2\beta^2 = (\alpha\beta)^2$

$\Rightarrow \quad \alpha^2\beta^2 = 7^2 = 49$

Two relations which will prove very useful are

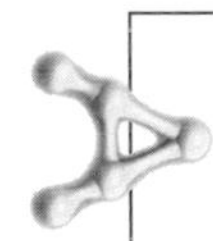

$$\alpha^2 + \beta^2 = (\alpha + \beta)^2 - 2\alpha\beta$$
$$\alpha^3 + \beta^3 = (\alpha + \beta)^3 - 3\alpha\beta(\alpha + \beta)$$

Notice how the expressions on the RHS contain combinations of just $\alpha + \beta$ and $\alpha\beta$.

These two results can be proved fairly easily:

$(\alpha + \beta)^2 = (\alpha + \beta)(\alpha + \beta) \Rightarrow (\alpha + \beta)^2 = \alpha^2 + 2\alpha\beta + \beta^2$

$\Rightarrow \alpha^2 + \beta^2 = (\alpha + \beta)^2 - 2\alpha\beta$ as required

and

$(\alpha + \beta)^3 = (\alpha + \beta)(\alpha + \beta)(\alpha + \beta)$

$\Rightarrow \quad (\alpha + \beta)^3 = (\alpha + \beta)(\alpha^2 + 2\alpha\beta + \beta^2)$

$\Rightarrow \quad (\alpha + \beta)^3 = \alpha^3 + 2\alpha^2\beta + \alpha\beta^2 + \alpha^2\beta + 2\alpha\beta^2 + \beta^3$

$\Rightarrow \quad (\alpha + \beta)^3 = \alpha^3 + 3\alpha^2\beta + 3\alpha\beta^2 + \beta^3$

$\Rightarrow \quad (\alpha + \beta)^3 = \alpha^3 + \beta^3 + 3\alpha\beta(\alpha + \beta)$ ← Take $3\alpha\beta$ as a factor.

$\Rightarrow \quad \alpha^3 + \beta^3 = (\alpha + \beta)^3 - 3\alpha\beta(\alpha + \beta)$ as required

Worked example 1.5

Given that $\alpha + \beta = 5$ and $\alpha\beta = -2$, find the values of:

(a) $\dfrac{\alpha}{\beta} + \dfrac{\beta}{\alpha}$, **(b)** $\alpha^3 + \beta^3$, **(c)** $\dfrac{1}{\alpha^2} + \dfrac{1}{\beta^2}$.

Solution

(a) $$\frac{\alpha}{\beta} + \frac{\beta}{\alpha} = \frac{\alpha^2 + \beta^2}{\alpha\beta} = \frac{(\alpha + \beta)^2 - 2\alpha\beta}{\alpha\beta}$$

Now it is in terms of $\alpha + \beta$ and $\alpha\beta$.

Substitute the known values of $\alpha + \beta$ and $\alpha\beta$

$$\Rightarrow \quad \frac{\alpha}{\beta} + \frac{\beta}{\alpha} = \frac{5^2 - 2(-2)}{-2} = \frac{25 + 4}{-2} = -\frac{29}{2}$$

(b) $\alpha^3 + \beta^3 = (\alpha + \beta)^3 - 3\alpha\beta(\alpha + \beta)$

$$\Rightarrow \quad \alpha 3 + \beta^3 = 5^3 - 3(-2)(5) = 125 + 30 = 155$$

(c) $$\frac{1}{\alpha^2} + \frac{1}{\beta^2} = \frac{\beta^2 + \alpha^2}{\alpha^2\beta^2} = \frac{(\alpha + \beta)^2 - 2\alpha\beta}{(\alpha\beta)^2}$$

$$\Rightarrow \quad \frac{1}{\alpha^2} + \frac{1}{\beta^2} = \frac{5^2 - 2(-2)}{(-2)^2} = \frac{25 + 4}{4} = \frac{29}{4}$$

Worked example 1.6

Given that $\alpha + \beta = 5$ and $\alpha\beta = \frac{2}{3}$, find the value of $(\alpha - \beta)^2$.

Solution

$$(\alpha - \beta)^2 = \alpha^2 - 2\alpha\beta + \beta^2 = \alpha^2 + \beta^2 - 2\alpha\beta$$

$$= (\alpha + \beta)^2 - 2\alpha\beta - 2\alpha\beta = (\alpha + \beta)^2 - 4\alpha\beta$$

$$\Rightarrow \quad (\alpha - \beta)^2 = 5^2 - 4\left(\tfrac{2}{3}\right) = 25 - \tfrac{8}{3} = 22\tfrac{1}{3}$$

Worked example 1.7

Write the expression $\dfrac{\alpha}{\beta^2} + \dfrac{\beta}{\alpha^2}$ in terms of $\alpha + \beta$ and $\alpha\beta$.

Solution

$$\frac{\alpha}{\beta^2} + \frac{\beta}{\alpha^2} = \frac{\alpha^3 + \beta^3}{\alpha^2\beta^2} = \frac{(\alpha + \beta)^3 - 3\alpha\beta(\alpha + \beta)}{(\alpha\beta)^2}$$

EXERCISE 1B

1 Write each of the following expressions in terms of $\alpha + \beta$ and $\alpha\beta$:

(a) $\frac{2}{\alpha} + \frac{2}{\beta}$ **(b)** $\frac{1}{\alpha^2\beta} + \frac{1}{\beta^2\alpha}$

(c) $\frac{\alpha}{3\beta} + \frac{\beta}{3\alpha}$ **(d)** $\alpha^2\beta + \beta^2\alpha$

(e) $(2\alpha - 1)(2\beta - 1)$ **(f)** $\frac{\alpha + 5}{\beta} + \frac{\beta + 5}{\alpha}$

2 Given that $\alpha + \beta = -3$ and $\alpha\beta = 9$, find the values of:

(a) $\alpha^3\beta + \beta^3\alpha$, **(b)** $\frac{\alpha}{\beta} + \frac{\beta}{\alpha}$.

3 Given that $\alpha + \beta = 4$ and $\alpha\beta = 10$, find the values of:

(a) $\alpha^2 + \beta^2$, **(b)** $\alpha^3 + \beta^3$.

4 Given that $\alpha + \beta = 7$ and $\alpha\beta = -2$, find the values of:

(a) $\frac{1}{\beta^2} + \frac{1}{\alpha^2}$, **(b)** $\frac{\alpha^2}{\beta} + \frac{\beta^2}{\alpha}$.

5 The roots of the quadratic equation $x^2 - 5x + 3 = 0$ are α and β.

(a) Write down the values of $\alpha + \beta$ and $\alpha\beta$.

(b) Hence find the values of:

(i) $(\alpha - 3)(\beta - 3)$, **(ii)** $\frac{\alpha}{\beta^2} + \frac{\beta}{\alpha^2}$.

6 The roots of the equation $x^2 - 4x + 3 = 0$ are α and β. Without solving the equation, find the value of:

(a) $\frac{1}{\alpha} + \frac{1}{\beta}$ **(b)** $\frac{\alpha}{\beta} + \frac{\beta}{\alpha}$

(c) $\alpha^2 + \beta^2$ **(d)** $\alpha^2\beta + \alpha\beta^2$

(e) $(\alpha - \beta)^2$ **(f)** $(\alpha + 1)(\beta + 1)$

7 The roots of the quadratic equation $x^2 + 4x + 1 = 0$ are α and β.

(a) Find the values of: **(i)** $\alpha + \beta$, **(ii)** $\alpha\beta$.

(b) Hence find the value of:

(i) $(\alpha^2 - \beta)(\beta^2 - \alpha)$, **(ii)** $\frac{\alpha + 2}{\beta + 2} + \frac{\beta + 2}{\alpha + 2}$.

1.3 Forming new equations with related roots

It is often possible to find a quadratic equation whose roots are related in some way to the roots of another given quadratic equation.

Worked example 1.8

The roots of the equation $2x^2 - 5x - 6 = 0$ are α and β.

Find a quadratic equation whose roots are $\frac{1}{\alpha}$ and $\frac{1}{\beta}$.

Solution

$$2x^2 - 5x - 6 = 0$$

$\Rightarrow$ The sum of roots, $\alpha + \beta = -\frac{b}{a} = -\frac{-5}{2} = \frac{5}{2}$.

The product of roots, $\alpha\beta = \frac{c}{a} = \frac{-6}{2} = -3$.

You would be able to write down the 'new' equation with roots $\frac{1}{\alpha}$ and $\frac{1}{\beta}$ if you could find the sum and the product of the new roots.

The sum of new roots, $\frac{1}{\alpha} + \frac{1}{\beta} = \frac{\alpha + \beta}{\alpha\beta}$

$$= \frac{\frac{5}{2}}{-3} = -\frac{5}{6}.$$

The product of new roots, $\frac{1}{\alpha} \times \frac{1}{\beta} = \frac{1}{\alpha\beta}$

$$= \frac{1}{-3} = -\frac{1}{3}.$$

Using $x^2 - (\text{sum of roots})x + (\text{product of roots}) = 0$

the equation with roots $\frac{1}{\alpha}$ and $\frac{1}{\beta}$ is

$$x^2 - \left(-\frac{5}{6}\right)x + \left(-\frac{1}{3}\right) = 0$$

$$\Rightarrow \quad x^2 + \frac{5}{6}x - \frac{1}{3} = 0$$

$$\Rightarrow \quad 6x^2 + 5x - 2 = 0.$$

Multiply through by 6 in order to obtain integer coefficients (because it is often required in an examination question).

The last example illustrates the basic method for forming new equations with roots that are related to the roots of a given equation:

1. Write down the sum of the roots, $\alpha + \beta$, and the product of the roots, $\alpha\beta$, of the given equation.
2. Find the sum and product of the new roots in terms of $\alpha + \beta$ and $\alpha\beta$.
3. Write down the new equation using
   $x^2 - (\text{sum of new roots})x + (\text{product of new roots}) = 0$.

Worked example 1.9

The roots of the equation $x^2 + 7x - 2 = 0$ are α and β.

Find the values of $\alpha^2 + \beta^2$ and $\alpha^2\beta^2$.

Hence, find a quadratic equation whose roots are α^2 and β^2.

Solution

From the equation $x^2 + 7x - 2 = 0$ you have

The sum of roots, $\alpha + \beta = -\dfrac{b}{a} = -\dfrac{7}{1} = -7.$

The product of roots, $\alpha\beta = \dfrac{c}{a} = \dfrac{-2}{1} = -2.$

Now, $\alpha^2 + \beta^2 = (\alpha + \beta)^2 - 2\alpha\beta$

$= (-7)^2 - 2(-2) = 49 + 4 = 53$

and $\alpha^2\beta^2 = (\alpha\beta)^2 = (-2)^2 = 4.$

Now since the roots of the new equation are α^2 and β^2, $\alpha^2 + \beta^2$ and $\alpha^2\beta^2$ are the sum and product of the new roots.

The required equation is

$x^2 -$ (sum of new roots)x + (product of new roots) $= 0$

$\Rightarrow$ $x^2 - 53x + 4 = 0.$

You could use any variable you choose. For instance, you could write $y^2 - 53y + 4 = 0$.

Worked example 1.10

The roots of the quadratic equation $x^2 + 5x - 3 = 0$ are α and β.

Find a quadratic equation whose roots are α^3 and β^3.

Solution

From the equation $x^2 + 5x - 3 = 0$ you have

The sum of roots, $\alpha + \beta = -\dfrac{b}{a} = -\dfrac{5}{1} = -5.$

The product of roots, $\alpha\beta = \dfrac{c}{a} = \dfrac{-3}{1} = -3.$

The new equation has roots α^3 and β^3.

The sum of new roots, $\alpha^3 + \beta^3 = (\alpha + \beta)^3 - 3\alpha\beta(\alpha + \beta)$

$=(-5)^3 - 3(-3)(-5)$

$= -125 - 45 = -170.$

The product of new roots, $\alpha^3\beta^3 = (\alpha\beta)^3 = (-3)^3 = -27.$

Using $x^2 -$ (sum of new roots)x + (product of new roots) $= 0$ the required equation is

$x^2 + 170x - 27 = 0.$

EXERCISE 1C

1 The roots of the equation $x^2 + 6x - 4 = 0$ are α and β.
Find a quadratic equation whose roots are α^2 and β^2.

2 The roots of the equation $x^2 + 3x - 5 = 0$ are α and β.
Find a quadratic equation whose roots are $\frac{2}{\alpha}$ and $\frac{2}{\beta}$.

3 The roots of the equation $x^2 - 9x + 5 = 0$ are α and β.
Find a quadratic equation whose roots are $\alpha + 2$ and $\beta + 2$.

4 Given that the roots of the equation $2x^2 + 5x - 3 = 0$ are α and β, find an equation with integer coefficients whose roots are $\alpha\beta^2$ and $\alpha^2\beta$.

5 Given that the roots of the equation $3x^2 - 6x + 1 = 0$ are α and β, find a quadratic equation with integer coefficients whose roots are α^3 and β^3.

6 The roots of the equation $2x^2 + 4x - 1 = 0$ are α and β.
Find a quadratic equation with integer coefficients whose roots are $\frac{\beta}{\alpha}$ and $\frac{\alpha}{\beta}$.

7 The roots of the equation $3x^2 - 6x - 2 = 0$ are α and β.

(a) Find the value of $\alpha^2 + \beta^2$.

(b) Find a quadratic equation whose roots are $\alpha^2 + 1$ and $\beta^2 + 1$.

8 The roots of the equation $2x^2 + 7x + 3 = 0$ are α and β.
Without solving this equation,

(a) find the value of $\alpha^3 + \beta^3$.

(b) Hence, find a quadratic equation with integer coefficients which has roots $\frac{\alpha}{\beta^2}$ and $\frac{\beta}{\alpha^2}$.

9 Given that α and β are the roots of the equation $5x^2 - 2x + 4 = 0$, find a quadratic equation with integer coefficients which has roots $\frac{1}{\alpha^2}$ and $\frac{1}{\beta^2}$.

10 The roots of the equation $x^2 + 4x - 6 = 0$ are α and β. Find an equation whose roots are $\alpha^2 + \beta$ and $\beta^2 + \alpha$.

1.4 Further examples

Sometimes the questions may require you to apply similar techniques to the previous section but without directing you to find the sum and product of the roots.

Worked example 1.11

The roots of the equation $x^2 - kx + 28 = 0$ are α and $\alpha + 3$.
Find the two possible values of k.

Solution

From $x^2 - kx + 28 = 0$ you have

The sum of roots, $\alpha + (\alpha + 3) = -\dfrac{-k}{1} \Rightarrow 2\alpha + 3 = k$ [1]

The product of roots, $\alpha(\alpha + 3) = \dfrac{28}{1} \Rightarrow \alpha^2 + 3\alpha = 28$ [2]

Equations [1] and [2] form a pair of simultaneous equations.

From [1] you have $\alpha = \dfrac{k-3}{2}$ and substituting this into [2] gives

$$\left(\frac{k-3}{2}\right)^2 + 3\left(\frac{k-3}{2}\right) = 28$$

$$\Rightarrow \quad \frac{k^2 - 6k + 9}{4} + \frac{3k - 9}{2} = 28$$

$$\Rightarrow \quad k^2 - 6k + 9 + 6k - 18 = 112$$

$$\Rightarrow \quad k^2 = 121$$

$\Rightarrow$ the two possible values for k are 11 and -11.

1.5 New equations by means of a substitution

Worked example 1.8 is reproduced below.

> The roots of the equation $2x^2 - 5x - 6 = 0$ are α and β.
>
> Find a quadratic equation whose roots are $\dfrac{1}{\alpha}$ and $\dfrac{1}{\beta}$.

Since the new equation has roots that are the reciprocals of the original equation, the new equation can be found very quickly by making the substitution $y = \dfrac{1}{x} \Rightarrow x = \dfrac{1}{y}$, which transforms

$2x^2 - 5x - 6 = 0$ into $\dfrac{2}{y^2} - \dfrac{5}{y} - 6 = 0.$

Multiplying throughout by y^2 gives

$$2 - 5y - 6y^2 = 0 \quad \text{or}$$
$$6y^2 + 5y - 2 = 0.$$

Check the working in Worked example 1.8 to see which is quicker.

Worked example 1.12

The roots of the equation $x^2 - 3x - 5 = 0$ are α and β.
Find quadratic equations with roots

(a) $\alpha - 3$ and $\beta - 3$, **(b)** α^2 and β^2.

You could solve this question using the sum and product of roots technique of the previous section if you prefer.

Solution

(a) Use the substitution $y = x - 3$ in $x^2 - 3x - 5 = 0$, since the roots in the new equation are 3 less than those in the original equation.

$y = x - 3 \quad \Rightarrow \quad x = y + 3$, so new equation is

$(y + 3)^2 - 3(y + 3) - 5 = 0$

or $\quad y^2 + 6y + 9 - 3y - 9 - 5 = 0$

The new equation is $y^2 + 3y - 5 = 0$.

If you prefer, you can write the new equation as $x^2 + 3x - 5 = 0$ or in terms of any other variable.

(b) This time you need to use the substitution $y = x^2$ in $x^2 - 3x - 5 = 0$ to eliminate x.

Hence, $y - 3x - 5 = 0$ or $y - 5 = 3x$.

Squaring both sides gives $(y - 5)^2 = 9x^2 = 9y$.

Hence, $y^2 - 10y + 25 = 9y$ or $y^2 - 19y + 25 = 0$.

Alternative solution

Using $\alpha + \beta = 3$ and $\alpha\beta = -5$

(a) Sum of new roots is

$$(\alpha - 3) + (\beta - 3) = \alpha + \beta - 6 = -3$$

Product of new roots is

$$\begin{aligned}(\alpha - 3)(\beta - 3) &= \alpha\beta - 3\alpha - 3\beta + 9 \\ &= \alpha\beta - 3(\alpha + \beta) + 9 = -5\end{aligned}$$

Hence new equation is $y^2 + 3y - 5 = 0$.

(b) Sum of new roots is

$$\begin{aligned}\alpha^2 + \beta^2 &= (\alpha + \beta)^2 - 2\alpha\beta \\ &= 9 + 10 = 19\end{aligned}$$

Product of new roots is

$$\alpha^2\beta^2 = (\alpha\beta)^2 = 25$$

Hence new equation is $y^2 - 19y + 25 = 0$.

You should see that the answers are the same whichever method you use. If the question directs you to use a particular substitution, e.g. 'Use the substitution $y = x^2$ to find a new equation whose roots are α^2 and β^2', you must use the first method. However, if the question is of the form 'Use the substitution $y = x^2$, or otherwise, to find a new equation whose roots are α^2 and β^2', or simply 'Find an equation whose roots are α^2 and β^2' you may use either technique.

Worked examination question 1.13

The roots of the quadratic equation $x^2 - 3x - 7 = 0$ are α and β.

(a) Write down the values of **(i)** $\alpha + \beta$, **(ii)** $\alpha\beta$.

(b) Find a quadratic equation with integer coefficients whose roots are $\dfrac{\alpha}{\beta}$ and $\dfrac{\beta}{\alpha}$.

Solution

(a) From the equation $x^2 - 3x - 7 = 0$ you have

The sum of roots, $\alpha + \beta = -\dfrac{b}{a} = -\dfrac{-3}{1} = 3$.

The product of roots, $\alpha\beta = \dfrac{c}{a} = \dfrac{-7}{1} = -7$.

(b) The new equation has roots $\dfrac{\alpha}{\beta}$ and $\dfrac{\beta}{\alpha}$.

The sum of new roots, $\dfrac{\alpha}{\beta} + \dfrac{\beta}{\alpha} = \dfrac{\alpha^2 + \beta^2}{\alpha\beta}$

$$= \frac{(\alpha + \beta)^2 - 2\alpha\beta}{\alpha\beta}$$

$$= \frac{3^2 - 2(-7)}{-7} = \frac{9 + 14}{-7} = -\frac{23}{7}$$

The product of new roots, $\dfrac{\alpha}{\beta} \times \dfrac{\beta}{\alpha} = \dfrac{\alpha\beta}{\alpha\beta} = 1$.

Using $x^2 - (\text{sum of new roots})x + (\text{product of new roots}) = 0$ the required equation is

$$x^2 - \left(-\frac{23}{7}\right)x + 1 = 0$$

$$\Rightarrow \quad x^2 + \frac{23}{7}x + 1 = 0$$

$$\Rightarrow \quad 7x^2 + 23x + 7 = 0$$

A common mistake is to forget to multiply throughout by 7 to obtain integer coefficients.

EXERCISE 1D

1 The roots of the equation $x^2 + 9x + k = 0$ are α and $\alpha + 1$. Find the value of k.

2 The roots of the quadratic equation $4x^2 + kx - 5 = 0$ are α and $\alpha - 3$. Find the two possible values of the constant k.

3 The roots of the quadratic equation $x^2 + 3x - 7 = 0$ are α and β. Find an equation with integer coefficients which has roots $\dfrac{3}{\alpha}$ and $\dfrac{3}{\beta}$.

4 The roots of the quadratic equation $2x^2 + 5x - 1 = 0$ are α and β. Find an equation with integer coefficients which has roots $\alpha - 2$ and $\beta - 2$.

5 The roots of the quadratic equation $x^2 - 2x - 5 = 0$ are α and β. Find a quadratic equation which has roots $\alpha^2 + 1$ and $\beta^2 + 1$.

6 The roots of the quadratic equation $x^2 + 5x - 7 = 0$ are α and β.

(a) Without solving the equation, find the values of

(i) $\alpha^2 + \beta^2$, **(ii)** $\alpha^3 + \beta^3$.

(b) Determine an equation with integer coefficients which has roots $\dfrac{\alpha^2}{\beta}$ and $\dfrac{\beta^2}{\alpha}$. [A]

7 The roots of the quadratic equation $3x^2 + 4x - 1 = 0$ are α and β.

(a) Without solving the equation, find the values of

(i) $\alpha^2 + \beta^2$, **(ii)** $\alpha^3\beta + \beta^3\alpha$.

(b) Determine a quadratic equation with integer coefficients which has roots $\alpha^3\beta$ and $\beta^3\alpha$. [A]

8 The roots of the quadratic equation $x^2 + 2x + 3 = 0$ are α and β.

(a) Without solving the equation:

(i) write down the value of $\alpha + \beta$ and the value of $\alpha\beta$,

(ii) show that $\alpha^3 + \beta^3 = 10$,

(iii) find the value of $\dfrac{1}{\alpha^3} + \dfrac{1}{\beta^3}$.

(b) Determine a quadratic equation with integer coefficients which has roots $\dfrac{1}{\alpha^3}$ and $\dfrac{1}{\beta^3}$. [A]

9 The roots of the quadratic equation $x^2 - 3x + 1 = 0$ are α and β.

(a) Without solving the equation:

(i) show that $\alpha^2 + \beta^2 = 7$,

(ii) find the value of $\alpha^3 + \beta^3$.

(b) **(i)** Show that $\alpha^4 + \beta^4 = (\alpha^2 + \beta^2)^2 - 2(\alpha\beta)^2$.

(ii) Hence, find the value of $\alpha^4 + \beta^4$.

(c) Determine a quadratic equation with integer coefficients which has roots $(\alpha^3 - \beta)$ and $(\beta^3 - \alpha)$. [A]

10 The roots of the quadratic equation $x^2 + 3x - 2 = 0$ are α and β.

(a) Write down the values of $\alpha + \beta$ and $\alpha\beta$.

(b) Without solving the equation, find the value of:

(i) $\dfrac{1}{\alpha^2} + \dfrac{1}{\beta^2}$, **(ii)** $\left(\alpha - \dfrac{3}{\beta^2}\right)\left(\beta - \dfrac{3}{\alpha^2}\right)$.

(c) Determine a quadratic equation with integer coefficients which has roots $\alpha - \dfrac{3}{\beta^2}$ and $\beta - \dfrac{3}{\alpha^2}$. [A]

Key point summary

1 When the quadratic equation $ax^2 + bx + c = 0$ has roots α and β: *p2*

- The sum of the roots, $\alpha + \beta = -\frac{b}{a}$;
- and the product of roots, $\alpha\beta = \frac{c}{a}$.

2 A quadratic equation can be expressed as $x^2 - (\text{sum of roots})x + (\text{product of roots}) = 0$. *p2*

3 Two useful results are: *p5*

$\alpha^2 + \beta^2 = (\alpha + \beta)^2 - 2\alpha\beta$

$\alpha^3 + \beta^3 = (\alpha + \beta)^3 - 3\alpha\beta(\alpha + \beta)$.

4 The basic method for forming new equations with roots that are related to the roots of a given equation is: *p8*

1. Write down the sum of the roots, $\alpha + \beta$, and the product of the roots, $\alpha\beta$, of the given equation.
2. Find the sum and product of the new roots in terms of $\alpha + \beta$ and $\alpha\beta$.
3. Write down the new equation using

$$x^2 - (\text{sum of new roots})x + (\text{product of new roots}) = 0$$

Test yourself — What to review

1 Find the sums and products of the roots of the following equations: *Section 1.1*

(a) $x^2 - 6x + 4 = 0$, **(b)** $2x^2 + x - 5 = 0$.

2 Write down the equations, with integer coefficients, where: *Section 1.3*

(a) sum of roots $= -4$, product of roots $= 7$,

(b) sum of roots $= \frac{5}{4}$, product of roots $= -\frac{1}{2}$.

3 The roots of the equation $x^2 + 5x - 6 = 0$ are α and β. Find the values of: *Section 1.2*

(a) $\alpha^2 + \beta^2$ **(b)** $\frac{\beta}{\alpha} + \frac{\alpha}{\beta}$.

4 Given that the roots of $3x^2 - 9x + 1 = 0$ are α and β, find a quadratic equation whose roots are $\frac{1}{\alpha\beta^2}$ and $\frac{1}{\alpha^2\beta}$. *Section 1.3*

Test yourself ANSWERS

1 **(a)** sum $= 6$, product $= 4$;
(b) sum $= -\frac{1}{2}$, product $= -\frac{5}{2}$.

2 **(a)** $x^2 + 4x + 7 = 0$; **(b)** $4x^2 - 5x - 2 = 0$.

3 **(a)** 37; **(b)** $-\frac{37}{6}$.

4 $x^2 - 27x + 27 = 0$.

CHAPTER 2

Complex numbers

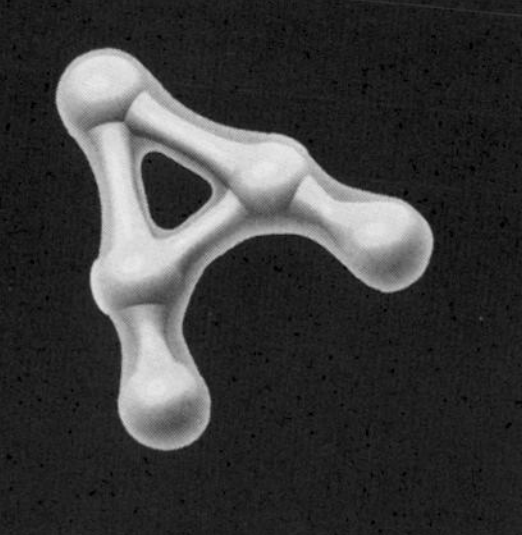

Learning objectives

After studying this chapter, you should be able to:

- understand what is meant by a complex number
- find complex roots of quadratic equations
- understand the term complex conjugate
- calculate sums, differences and products of complex numbers
- compare real and imaginary parts of complex numbers

2.1 The historical background to complex numbers

The history of complex numbers goes back to the ancient Greeks who decided, even though they were somewhat puzzled by the fact, that no real number existed that satisfies the equation $x^2 = -1$.

Diophantus, around 275 AD, attempted to solve what seems a reasonable problem, namely:

> Find the sides of a right-angled triangle with perimeter 12 units and area 7 squared units.

Let $AB = x$ and $AC = h$ as shown in the diagram.

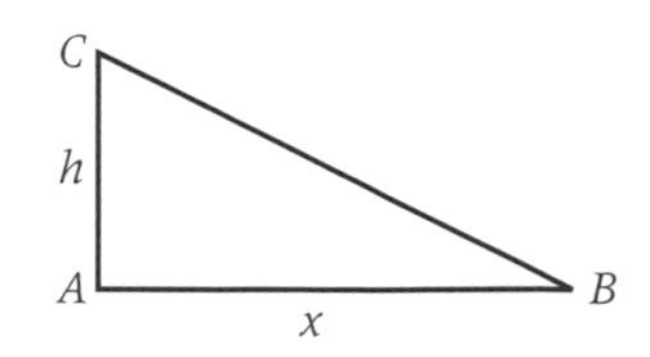

The perimeter is $x + h + \sqrt{(x^2 + h^2)}$ and the area is $\frac{1}{2}xh$.

Therefore, $x + h + \sqrt{(x^2 + h^2)} = 12$ and $\frac{1}{2}xh = 7$.

Writing $\sqrt{(x^2 + h^2)} = 12 - (x + h)$ and squaring both sides gives

$$x^2 + h^2 = 12^2 - 24(x + h) + (x^2 + 2hx + h^2)$$

and using $xh = 14$ gives

$$0 = 144 - 24(x + h) + 28$$

or $\quad 0 = 36 - 6(x + h) + 7.$

Multiplying throughout by x gives

$$0 = 36x - 6x^2 - 6hx + 7x$$

Since $xh = 14$.

or $\quad 0 = 43x - 6x^2 - 84.$

which can be rewritten as the quadratic equation

$$6x^2 - 43x + 84 = 0.$$

Check that $43^2 - 4 \times 6 \times 84$ is negative and therefore the quadratic equation has no real solutions.

However, as you learned in C1 chapter 4, when the discriminant $b^2 - 4ac$ of a quadratic expression is negative, the quadratic equation has no real solutions.

Worked example 2.1

The roots of the quadratic equation $x^2 - 3x + 7 = 0$ are α and β.

Without solving the equation, write down the values of $\alpha + \beta$ and $\alpha\beta$ and hence find the value of $\alpha^2 + \beta^2$.

What can you deduce about α and β?

Solution

$\alpha + \beta = 3$ and $\alpha\beta = 7$.

Hence, $\alpha^2 + \beta^2 = (\alpha + \beta)^2 - 2\alpha\beta = 9 - 14 = -5$.

It is not possible for the sum of the squares of two real numbers to be negative. Hence, α and β cannot both be real.

You will see later in this chapter that, when this is the situation, we say that α and β are complex numbers.

2.2 The imaginary number i

An imaginary number denoted by the symbol i is introduced so we can represent numbers that are not real. It is defined to have the special property that when you square it you get -1.

If you try to find the square root of -1 on your calculator, it is likely to give you an error message. Try it and see.

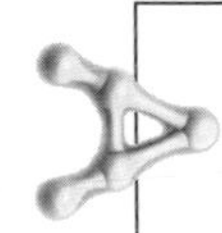

The square root of -1 is not real.
Because it is imaginary, it is denoted by i where $i^2 = -1$.

This means that when you square 3i you get

$$(3i)^2 = 3^2 \times i^2 = 9 \times -1 = -9.$$

You should now be able to see how we can find the square roots of negative numbers using i.

Worked example 2.2

Simplify each of the following powers of i:

(a) i^3, **(b)** i^4, **(c)** i^7.

Solution

(a) Since $i^3 = i^2 \times i$ and using the fact that $i^2 = -1$, you have $i^3 = -1 \times i = -i$.

(b) $i^4 = i^2 \times i^2 = -1 \times -1 = +1$.
So $i^4 = 1$.

(c) $i^7 = i^4 \times i^3 = +1 \times -i = -i$, using the previous results.
Hence, $i^7 = -i$.

When evaluating powers of i, it is very useful to know that $i^4 = 1$. So, for example

$$i^{29} = (i^4)^7 \times i = 1 \times i = i.$$

Since $i^2 = -1$, it follows that $i^4 = 1$.
The powers of i form a periodic cycle of the form $i, -1, -i, +1, \ldots$, so that $i^5 = i$, $i^6 = -1$, etc.

By introducing the symbol i it is possible to solve equations of the form $x^2 = -k$, where k is positive.

Since $(-i)^2 = (-1)^2 \times i^2 = +1 \times -1 = -1$, it follows that:

The equation $x^2 = -1$ has the two solutions:
$x = i$ and $x = -i$.

Worked example 2.3

Solve the following equations, giving your answers in terms of i:

(a) $x^2 = -4$, **(b)** $y^2 + 81 = 0$, **(c)** $z^2 = -12$.

Solution

(a) Since the equation $x^2 = -1$ has the two solutions $x = i$ and $x = -i$, and since $4 = 2^2$ it follows that $x^2 = -4$ has solutions $x = 2i$ and $x = -2i$.

(b) Rewriting the equation in the form $y^2 = -81 = -1 \times 9^2$, the two solutions are $y = 9i$ and $y = -9i$.

(c) $z^2 = -12 = -1 \times 12$.
Recall that, using surds, $\sqrt{12} = \sqrt{4} \times \sqrt{3} = 2\sqrt{3}$.
Therefore the equation has solutions $z = 2\sqrt{3}i$ and $z = -2\sqrt{3}i$.

2

EXERCISE 2A

1 Simplify each of the following:

(a) i^5, **(b)** i^6, **(c)** i^9, **(d)** i^{27},
(e) $(-i)^3$, **(f)** $(-i)^7$, **(g)** $(-i)^{10}$.

2 Simplify:

(a) $(2i)^3$, **(b)** $(3i)^4$, **(c)** $(7i)^2$,
(d) $(-2i)^2$, **(e)** $(-3i)^3$, **(f)** $(-2i)^5$.

3 Solve each of the following equations, giving your answers in terms of i:

(a) $x^2 = -9$, **(b)** $x^2 = -100$, **(c)** $x^2 = -49$,
(d) $x^2 + 1 = 0$, **(e)** $x^2 + 121 = 0$, **(f)** $x^2 + 64 = 0$,
(g) $x^2 + n^2 = 0$, where n is a positive integer.

4 Find the exact solutions of each of the following equations, giving your answers in terms of i:

(a) $x^2 = -5$, **(b)** $x^2 = -3$, **(c)** $x^2 = -8$,
(d) $x^2 + 20 = 0$, **(e)** $x^2 + 18 = 0$, **(f)** $x^2 + 48 = 0$.

5 In 1545, the mathematician Cardano tried to solve the problem of finding two numbers, a and b, whose sum is 10 and whose product is 40.

(a) Show that a must satisfy the equation $a^2 - 10a + 40 = 0$.

(b) What can you deduce about the numbers a and b?

6 A right-angled triangle has area A cm^2 and perimeter P cm. A side other than the hypotenuse has length x cm.
Form a quadratic equation in x in each of the following cases:

(a) $A = 6$, $P = 12$, **(b)** $A = 3$, $P = 8$, **(c)** $A = 30$, $P = 30$.

Hence, for each case above, find the possible values of x whenever real solutions exist.

2.3 Complex numbers and complex conjugates

Whereas a number such as 3i is said to be imaginary, a number such as $3 + 5i$ is said to be complex. It consists of the real number 3 added to the imaginary number 5i.

A number of the form $p + qi$, where p and q are real numbers and $i^2 = -1$, is called a **complex number**.

Imaginary numbers are in fact a subset of the complex numbers, since any imaginary number can be written as $0 + ki$, where k is real.

The two complex numbers $3 + 5i$ and $3 - 5i$ are said to be **complex conjugates**.

You met the idea of *real* conjugates when dealing with surds in C1 where, for example, $2 + \sqrt{3}$ and $2 - \sqrt{3}$ were said to be conjugates.

The complex conjugate of $3 + 5i$ is $3 - 5i$ and the complex conjugate of $3 - 5i$ is $3 + 5i$. Consequently, the numbers $3 + 5i$ and $3 - 5i$ are often referred to as a conjugate pair.

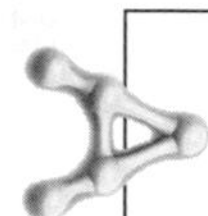

The complex conjugate of $p + qi$ is $p - qi$, where p and q are real numbers.
The conjugate of z is denoted by z^*.

Hence, if $z = -7 + 2i$, you can immediately write down that $z^* = -7 - 2i$.

When $z = 3 - 4i$, for example, $z^* = 3 + 4i$.

Worked example 2.4

Write down the complex conjugates of

(a) $-2 + 3i$, **(b)** $8 - 9i$, and **(c)** $2 + \sqrt{3}i$.

Solution

(a) The complex conjugate of $-2 + 3i$ is $-2 - 3i$;

Notice that the real part remains as -2.

(b) The complex conjugate of $8 - 9i$ is $8 + 9i$;

(c) The complex conjugate of $2 + \sqrt{3}i$ is $2 - \sqrt{3}i$.

2.4 Combining complex numbers

The operations of addition, subtraction and multiplication using complex numbers are very similar to the way you perform arithmetic with real numbers.

In algebra, you can simplify expressions such as $(3 + 4x) - 2(5 + x)$.

$$(3 + 4x) - 2(5 + x) = 3 + 4x - 10 - 2x = -7 + 2x.$$

Similarly, with complex numbers, you can simplify $(3 + 4i) - 2(5 + i)$.

$$(3 + 4i) - 2(5 + i) = 3 + 4i - 10 - 2i = -7 + 2i.$$

You can multiply complex numbers in the same way that you multiply out brackets in algebra.

$$(4 + 5x)(3 - 2x) = 12 - 8x + 15x - 10x^2 = 12 + 7x - 10x^2.$$

Similarly,

$$\begin{aligned}(4 + 5i)(3 - 2i) &= 12 - 8i + 15i - 10i^2 = 12 + 7i - 10 \times (-1) \\ &= 22 + 7i\end{aligned}$$

Since $i^2 = -1$.

The letter z is often used to represent a complex number. If a question uses more than one complex number, it is common to use subscripts and to write the numbers as z_1, z_2, z_3, etc.

Worked example 2.5

The complex numbers z_1 and z_2 are given by $z_1 = 3 - 5i$ and $z_2 = -1 + 4i$.

Find: **(a)** $2z_1 + 5z_2$ **(b)** $3z_1 - 4z_2$
(c) z_1z_2 **(d)** z_1^2

Solution

(a) $2z_1 + 5z_2 = 2(3 - 5i) + 5(-1 + 4i) = 6 - 10i - 5 + 20i$
$= 1 + 10i$

(b) $3z_1 - 4z_2 = 3(3 - 5i) - 4(-1 + 4i) = 9 - 15i + 4 - 16i$
$= 13 - 31i$

(c) $z_1z_2 = (3 - 5i)(-1 + 4i) = -3 + 12i + 5i - 20i^2$
$= -3 + 17i + 20$
$= 17 + 17i$

(d) $z_1^2 = (3 - 5i)(3 - 5i) = 9 - 30i + 25i^2 = 9 - 30i - 25$
$= -16 - 30i$

You will find that when you are solving a quadratic equation with real coefficients, if one of the roots is $3 + 5i$, for example, then the other root is $3 - 5i$.

Worked example 2.6

Find the quadratic equation with roots $3 + 5i$ and $3 - 5i$.

Solution

The sum of the roots is $(3 + 5i) + (3 - 5i) = 6 + 0i = 6$.

The product of the roots is $(3 + 5i)(3 - 5i)$

$$= 9 + 15i - 15i - 5^2i^2 = 9 - 25 \times (-1) = 9 + 25$$
$$= 34.$$

Hence, the quadratic equation is $x^2 - 6x + 34 = 0$.

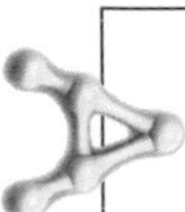

When a quadratic equation with real coefficients has complex roots, these roots are always a pair of complex conjugates.

EXERCISE 2B

1 Find the complex conjugate of each of the following:

(a) $3 - \mathrm{i}$, **(b)** $2 + 6\mathrm{i}$, **(c)** $-3 - 8\mathrm{i}$, **(d)** $-7 + 5\mathrm{i}$,
(e) $3 + \sqrt{2}\mathrm{i}$, **(f)** $4 - \sqrt{3}\mathrm{i}$, **(g)** $-1 - \frac{1}{3}\mathrm{i}$

2 Simplify each of the following:

(a) $(3 + \mathrm{i}) + (5 - 2\mathrm{i})$, **(b)** $(3 + \mathrm{i}) - (5 - 2\mathrm{i})$,
(c) $3(3 - 5\mathrm{i}) + 4(1 + 6\mathrm{i})$, **(d)** $3(3 - 5\mathrm{i}) - 4(1 + 6\mathrm{i})$,
(e) $4(8 - 5\mathrm{i}) - 5(1 - 4\mathrm{i})$, **(f)** $6(3 - 4\mathrm{i}) - 2(9 - 6\mathrm{i})$.

3 Simplify each of the following:

(a) $(4 + \mathrm{i})(7 - 2\mathrm{i})$, **(b)** $(3 + 4\mathrm{i})(5 - 3\mathrm{i})$,
(c) $(7 - 5\mathrm{i})(5 + 6\mathrm{i})$, **(d)** $3(3 - 5\mathrm{i})(4 + 3\mathrm{i})$,
(e) $(8 - 5\mathrm{i})(8 - 5\mathrm{i})$, **(f)** $(3 - 4\mathrm{i})^2$.

4 Find the square of each of the following complex numbers:

(a) $3 - \mathrm{i}$, **(b)** $2 + 6\mathrm{i}$,
(c) $-3 - 8\mathrm{i}$, **(d)** $-7 + 5\mathrm{i}$,
(e) $3 + \sqrt{2}\mathrm{i}$, **(f)** $4 - \sqrt{3}\mathrm{i}$.

5 The complex numbers z_1 and z_2 are given by $z_1 = 2 - 3\mathrm{i}$ and $z_2 = -3 + 5\mathrm{i}$.

Find: **(a)** $5z_1 + 3z_2$ **(b)** $3z_1 - 4z_2$ **(c)** z_1z_2

6 Given that z^* is the conjugate of z, find the values of

(i) $z + z^*$, **(ii)** $z - z^*$, **(iii)** zz^*

for each of the following values of z:

(a) $2 + 3\mathrm{i}$, **(b)** $-4 + 2\mathrm{i}$, **(c)** $-5 - 3\mathrm{i}$, **(d)** $6 - 5\mathrm{i}$,
(e) $x + y\mathrm{i}$, where x and y are real.

7 Find the value of the real constant p so that $(3 + 2\mathrm{i})(4 - \mathrm{i}) + p$ is purely imaginary.

8 Find the value of the real constant q so that $(2 + 5\mathrm{i})(4 - 3\mathrm{i}) + q\mathrm{i}$ is real.

9 **(a)** Find $(1 + \mathrm{i})(2 - 3\mathrm{i})$.

(b) Hence, simplify $(1 + \mathrm{i})(2 - 3\mathrm{i})(5 + \mathrm{i})$.

10 **(a)** Find $(2 + \mathrm{i})^2$.

(b) Hence, find: **(i)** $(2 + \mathrm{i})^3$, **(ii)** $(2 + \mathrm{i})^4$.

2.5 Complex roots of quadratic equations

Perhaps the easiest way to solve quadratic equations with complex roots is to complete the square.

Recall that the quadratic equation $ax^2 + bx + c = 0$ will not have real roots when $b^2 - 4ac < 0$.

Worked example 2.7

Solve the following quadratic equations by completing the square.

(a) $x^2 - 4x + 13 = 0$,

(b) $x^2 + 6x + 25 = 0$,

(c) $x^2 - 2x + 7 = 0$.

Solution

(a) $x^2 - 4x + 13 = (x - 2)^2 + 9$

$$x^2 - 4x + 13 = 0 \Rightarrow (x - 2)^2 + 9 = 0 \Rightarrow (x - 2)^2 = -9$$
$$\Rightarrow (x - 2) = \pm 3\text{i} \Rightarrow x = 2 \pm 3\text{i}$$

Hence, the roots are $2 + 3\text{i}$ and $2 - 3\text{i}$.

Notice that the solutions are complex conjugates.

(b) $x^2 + 6x + 25 = (x + 3)^2 + 16$

$$x^2 + 6x + 25 = 0 \Rightarrow (x + 3)^2 + 16 = 0 \Rightarrow (x + 3)^2 = -16$$
$$\Rightarrow (x + 3) = \pm 4\text{i} \Rightarrow x = -3 \pm 4\text{i}$$

Hence, the roots are $-3 + 4\text{i}$ and $-3 - 4\text{i}$.

You can check your answers by finding the sum of the roots and the product of the roots.

(c) $x^2 - 2x + 7 = (x - 1)^2 + 6$

$$x^2 - 2x + 7 = 0 \Rightarrow (x - 1)^2 + 6 = 0 \Rightarrow (x - 1)^2 = -6$$
$$\Rightarrow (x - 1) = \pm\sqrt{6}\text{i} \Rightarrow x = 1 \pm \sqrt{6}\text{i}$$

Hence, the roots are $1 + \sqrt{6}\text{i}$ and $1 - \sqrt{6}\text{i}$.

Worked example 2.8

Solve the following quadratic equations by using the quadratic equation formula.

(a) $x^2 - 10x + 26 = 0$, **(b)** $2x^2 + 4x + 5 = 0$.

An alternative method is to use the quadratic equation formula, but this is sometimes more difficult than completing the square.

Solution

(a) Using $x = \dfrac{-b \pm \sqrt{b^2 - 4ac}}{2a}$ with $a = 1$, $b = -10$, $c = 26$:

$$x = \frac{10 \pm \sqrt{100 - 104}}{2}$$

Therefore, $x = \dfrac{10 \pm 2\text{i}}{2} = 5 \pm \text{i}$.

The solutions are $x = 5 + \text{i}$ and $x = 5 - \text{i}$.

(b) Using $x = \dfrac{-b \pm \sqrt{b^2 - 4ac}}{2a}$ with $a = 2$, $b = 4$, $c = 5$:

$$x = \frac{-4 \pm \sqrt{16 - 40}}{4}$$

Therefore, $x = \dfrac{-4 \pm \sqrt{24}\text{i}}{4} = \dfrac{-4 \pm 2\sqrt{6}\text{i}}{4} = -1 \pm \dfrac{\sqrt{6}}{2}\text{i}$.

The solutions are $x = -1 + \dfrac{\sqrt{6}}{2}\text{i}$ and $x = -1 - \dfrac{\sqrt{6}}{2}\text{i}$.

EXERCISE 2C

Find the complex roots of each of the following equations:

1 $x^2 - 2x + 5 = 0$

2 $x^2 + 4x + 13 = 0$

3 $x^2 - 2x + 10 = 0$

4 $x^2 - 6x + 25 = 0$

5 $x^2 - 8x + 20 = 0$

6 $x^2 + 4x + 5 = 0$

7 $x^2 - 12x + 40 = 0$

8 $x^2 + 2x + 50 = 0$

9 $x^2 + 8x + 17 = 0$

10 $x^2 - 10x + 34 = 0$

11 $2x^2 - 2x + 5 = 0$

12 $9x^2 + 6x + 10 = 0$

13 $4x^2 - 8x + 5 = 0$

14 $5x^2 - 6x + 5 = 0$

15 $13x^2 + 10x + 13 = 0$

16 $x^2 - 2x + 4 = 0$

17 $x^2 + 4x + 9 = 0$

18 $x^2 - 6x + 16 = 0$

19 $x^2 + 8x + 19 = 0$

20 $x^2 - x + 1 = 0$

2.6 Equating real and imaginary parts

The complex number $3 + 4\text{i}$ is said to have

real part equal to 3

and imaginary part equal to 4.

A common mistake is to say that the imaginary part is 4i, when it should be simply 4.

In general, when $z = p + q\text{i}$, where p and q are real numbers,
- the real part of z is p, and
- the imaginary part of z is q.

When an equation involves complex numbers, it is in fact TWO equations since the real parts must be the same on both sides of the equation and the imaginary parts must be equal as well.

Worked example 2.9

Find the value of each of the real constants a and b such that $(a + 2\text{i})^2 = b + 12\text{i}$.

Solution

The left-hand side can be multiplied out to give

$$a^2 + 4a\text{i} + 4\text{i}^2 = a^2 - 4 + 4a\text{i}.$$

Hence, $a^2 - 4 + 4a\text{i} = b + 12\text{i}$.

Equating real parts gives $\quad a^2 - 4 = b \quad$ [1]

and equating imaginary parts gives $\quad 4a = 12 \quad$ [2]

From [2], $a = 3$.

Substituting into [1] gives $9 - 4 = b$, so $b = 5$.

Worked example 2.10

Find the complex number z which satisfies the equation $2z + 3z^* = 10 + 4i$.

Solution

Let $z = x + yi$, where x and y are real.

Therefore, the conjugate $z^* = x - yi$.

$$2z + 3z^* = 2(x + yi) + 3(x - yi) = 2x + 2yi + 3x - 3yi = 5x - yi$$

Hence, $5x - yi = 10 + 4i$.

Equating real parts gives $5x = 10 \Rightarrow x = 2$.

Equating imaginary parts gives $-yi = 4i \Rightarrow y = -4$.

Hence, z is the complex number $2 - 4i$.

Worked example 2.11

The complex number $1 + 3i$ is a root of the quadratic equation $z^2 - (2 + 5i)z + p + iq = 0$, where p and q are real.

(a) Find the values of p and q.

(b) By considering the sum of the roots, find the second root of the quadratic equation. Why are the two roots not complex conjugates?

Solution

(a) Substituting $z = 1 + 3i$ into $z^2 - (2 + 5i)z + p + iq = 0$ gives

$$(1 + 3i)^2 - (2 + 5i)(1 + 3i) + p + iq = 0.$$

Expanding gives $1 + 6i - 9 - (2 + 5i + 6i - 15) + p + iq = 0$

or $-8 + 6i + 13 - 11i + p + iq = 0$.

Equating real parts gives $p = -5$.

Equating imaginary parts gives $q = 5$.

(b) The sum of the roots is $2 + 5i$ and one root is known to be $1 + 3i$. The other root must be $1 + 2i$.

The roots are not conjugates because the quadratic equation did not have real coefficients.

An interesting check can now be made of the values of p and q.

The product of the roots is $(1 + 3i)(1 + 2i) = 1 + 5i - 6 = -5 + 5i$, which confirms that the values of p and q were correct.

Worked example 2.12

Find the square roots of the complex number $-5 + 12\text{i}$.

Solution

You need to find possible complex numbers z such that $z^2 = -5 + 12\text{i}$.

Let $z = x + y\text{i}$, where x and y are real.

$$z^2 = (x + y\text{i})^2 = x^2 + 2xy\text{i} - y^2 = -5 + 12\text{i}.$$

Equating real and imaginary parts gives

$$x^2 - y^2 = -5 \qquad [1]$$

and $\quad 2xy = 12 \;$ or $\; xy = 6 \quad [2]$

Substitute $y = \dfrac{6}{x}$ into [1]

$$x^2 - \frac{36}{x^2} = -5 \Rightarrow x^4 - 36 = -5x^2$$

or $\quad x^4 + 5x^2 - 36 = 0 \Rightarrow (x^2 + 9)(x^2 - 4) = 0$

So either $x^2 = -9$ or $x^2 = 4$.

But since x is real, the only possibility is that $x^2 = 4$. This gives two values for x, $x = 2$ or $x = -2$.

Since $y = \dfrac{6}{x}$, $x = 2 \Rightarrow y = 3$ and $x = -2 \Rightarrow y = -3$.

Hence, the two square roots of $-5 + 12\text{i}$ are $2 + 3\text{i}$ and $-2 - 3\text{i}$.

EXERCISE 2D

1 Find the value of each of the real constants a and b such that $(a + 3\text{i})^2 = 8b + 30\text{i}$.

2 Find the value of each of the real constants p and q such that $(3 + 4\text{i})(p + 2\text{i}) = q + 26\text{i}$.

3 Find the value of each of the real constants t and u such that $(2 + 2\text{i})^2(t + 3\text{i}) = u + 32\text{i}$.

4 Find the complex number z so that $3z + 4z^* = 28 + \text{i}$.

5 Find the complex number z so that $z + 3z^* = 12 - 8\text{i}$.

6 Find the complex number z so that $z - 4\text{i}z^* + 2 + 7\text{i} = 0$.

7 Find all possible values of z such that $z + z^* = 6$.

8 Find the square roots of each of the following complex numbers:

(a) $-7 + 24\text{i}$, **(b)** $35 - 12\text{i}$, **(c)** $21 + 20\text{i}$.

9 The complex number $2 + 5\text{i}$ is a root of the quadratic equation $z^2 - (3 + 7\text{i})z + p + \text{i}q = 0$, where p and q are real.

(a) Find the values of p and q.

(b) By considering the sum of the roots, find the second root of the quadratic equation. Why are the two roots not complex conjugates?

10 (a) Find the square roots of $-3 + 4\text{i}$.

(b) Hence, or otherwise, find the two complex roots of the quadratic equation $z^2 + 2(1 + \text{i})z + 3 - 2\text{i} = 0$.

Key point summary

1 The square root of -1 is not real. Because it is imaginary, it is denoted by i where $\text{i}^2 = -1$. *p17*

2 Since $\text{i}^2 = -1$, it follows that $\text{i}^4 = 1$. *p18*
The powers of i form a periodic cycle of the form $\text{i}, -1, -\text{i}, +1, \ldots$, so that $\text{i}^5 = \text{i}$, $\text{i}^6 = -1$, etc.

3 The equation $x^2 = -1$ has the two solutions: *p18*
$x = \text{i}$ and $x = -\text{i}$.

4 A number of the form $p + q\text{i}$, where p and q are real numbers and $\text{i}^2 = -1$, is called a complex number. *p19*

5 The complex conjugate of $p + q\text{i}$ is $p - q\text{i}$, where p and q are real numbers. *p20*
The conjugate of z is denoted by z^*.

6 When a quadratic equation with real coefficients has complex roots, these roots are always a pair of complex conjugates. *p21*

7 In general, when $z = p + q\text{i}$, where p and q are real numbers, *p24*

- the real part of z is p, and
- the imaginary part of z is q.

Test yourself	What to review
1 Simplify: **(a)** i^7, **(b)** i^{34}.	*Section 2.2*
2 Solve the equation $x^2 + 16 = 0$.	*Section 2.2*
3 State the complex conjugates of: **(a)** $-2 + 7i$, **(b)** $5 - 4i$.	*Section 2.3*
4 Simplify: **(a)** $(-2 + 3i) - 4(-1 + i)$, **(b)** $(5 - 4i)(3 - 2i)$.	*Section 2.4*
5 Find the complex roots of the equation $x^2 + 4x + 29 = 0$.	*Section 2.5*
6 Find the complex number z that satisfies the equation $z + 3z^* = 4 + 8i$.	*Section 2.6*

Test yourself ANSWERS

1 **(a)** $-i$; **(b)** -1.

2 $-4i, 4i$.

3 **(a)** $-2 - 7i$; **(b)** $5 + 4i$.

4 **(a)** $2 - i$; **(b)** $7 - 22i$.

5 $-2 + 5i, -2 - 5i$.

6 $1 - 4i$.

CHAPTER 3

Inequalities

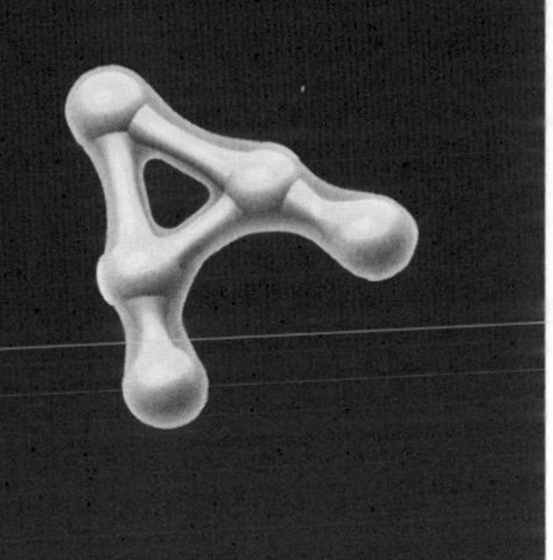

Learning objectives

After studying this chapter, you should be able to:

- use sign diagrams to solve inequalities
- solve inequalities involving rational expressions.

This chapter covers the various techniques for solving inequalities involving linear rational expressions.

3.1 Introduction

You solve inequalities in a similar way to solving equations. However, instead of the = sign, you manipulate one of the following signs:

$>$ greater than

$\geqslant$ greater than or equal to

$<$ less than

$\leqslant$ less than or equal to

You have learned to solve simple linear and quadratic inequalities in the C1 module. An example is given below to remind you.

Worked example 3.1

(a) Solve the linear inequality $2(3 - 5x) < 6 - 2(3x + 4)$.

(b) Solve the quadratic inequality $x(x + 2) > 15$.

Solution

(a) $2(3 - 5x) < 6 - 2(3x + 4)$

$\Rightarrow \quad 6 - 10x < 6 - 6x - 8$

$\Rightarrow \quad -10x + 6x < 6 - 8 - 6$ ← Collect the *x*s on the left and the numbers on the right.

$\Rightarrow \quad -4x < -8$ ← Simplify.

$\Rightarrow \quad x > 2$ ← Divide by -4, but notice how the inequality is **reversed**.

(b) $x(x + 2) > 15$

$\Rightarrow \quad x^2 + 2x - 15 > 0$

$\Rightarrow \quad (x + 5)(x - 3) > 0$

Multiply out, collect terms and factorise.

Let $f(x) = (x + 5)(x - 3)$.

We introduced the term **critical values** in C1. The critical values for this particular expression are $x = -5$ and $x = 3$.

This is because these values make each of the brackets in turn equal to 0.

Consider a number line with these critical values marked.

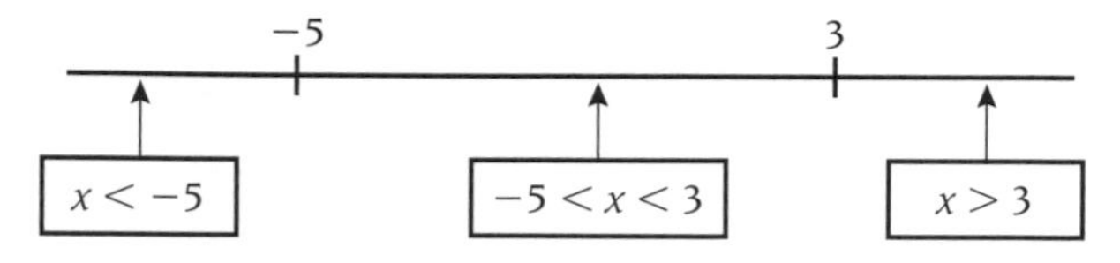

$x < -5$, choose $x = -6$
$f(-6) = (-6 + 5)(-6 - 3)$
$= -1 \times -9 = 9$

$-5 < x < 3$, choose $x = 0$
$f(0) = (0 + 5)(0 - 3)$
$= 5 \times -3 = -15$

$x > 3$, choose $x = 4$
$f(4) = (4 + 5)(4 - 3)$
$= 9 \times 1 = 9$

The line has been cut into three separate regions:
$x < -5$, $-5 < x < 3$ and $x > 3$.

The method consists of choosing a value in each of these regions and calculating the value of the expression $f(x)$.

You can then draw a sign diagram for $f(x) = (x + 5)(x - 3)$.

There are two regions where $f(x) > 0$.

The solution is therefore in two parts: $x < -5$, $x > 3$.

You need to know when the expression is positive.

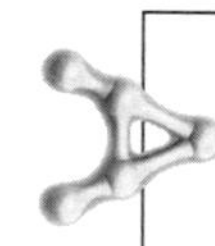

There are important differences between solving equations and solving inequalities.

1. An inequality will have a **range** of values as its solution.
2. Whenever you multiply or divide an inequality by a **negative** number you must also **reverse** the inequality sign.

3.2 Inequalities involving rational expressions

Firstly, what does the term 'rational expression' mean?

Quite simply, a rational expression is one that contains an algebraic fraction such as $\dfrac{x + 7}{x - 2}$.

Consider the equation $\dfrac{x + 7}{x - 2} = 3$.

You can multiply both sides by $(x - 2)$ as this eliminates the fraction.

$\Rightarrow \quad (x + 7) = 3(x - 2) = 3x - 6$

$\Rightarrow \quad 13 = 2x$

$\Rightarrow \quad x = 6\frac{1}{2}$

How do you solve an inequality involving a rational expression such as $\frac{x+7}{x-2} < 3$?

One approach is to make sure you are multiplying throughout by a quantity such as $(x-2)^2$ since a squared quantity is never negative.

Hence, there is no need to reverse the inequality.

Warning: You cannot simply multiply both sides by $(x-2)$ since you don't know whether its value is positive or negative. If it were negative then you would have to reverse the inequality sign.
If $(x-2) > 0$ you would keep the inequality sign the same.

3

3.3 Multiplying both sides by the square of the denominator

Worked example 3.2

Solve the inequality $\frac{x+7}{x-2} < 3$.

Solution

$$\frac{x+7}{x-2} < 3$$

Multiplying both sides by $(x-2)^2$ gives

$$\frac{(x+7)(x-2)^2}{(x-2)} < 3(x-2)^2$$

$$\Rightarrow \quad (x+7)(x-2) < 3(x-2)^2$$

$$\Rightarrow \quad (x+7)(x-2) - 3(x-2)^2 < 0$$

Take all terms to the left-hand side.

$$\Rightarrow \quad (x-2)[(x+7) - 3(x-2)] < 0$$

Take $(x-2)$ as a common factor.

$$\Rightarrow \quad (x-2)(13-2x) < 0$$

Simplify the second bracket.

You then continue in the usual way for solving quadratic inequalities.

Let $\quad f(x) = (x-2)(13-2x)$

The critical values are $x = 2$ and $x = \frac{13}{2}$.

A sign diagram for $(x-2)(13-2x)$ is given below

	2		$\frac{13}{2}$	
–		+		–

The critical values cut the line into three regions:
$x < 2$, $2 < x < \frac{13}{2}$ and $x > \frac{13}{2}$.

You are trying to solve $(x-2)(13-2x) < 0$.

There are two regions which gave a negative value for $f(x)$.
The solution is in two parts: $x < 2$, $x > \frac{13}{2}$.

When solving quadratic/cubic/higher order inequalities you must consider the **critical values**.

You calculate the value of $f(x)$ in each of the regions on the number line created by the critical values and produce a sign diagram.

Worked example 3.3

Solve the inequality $\frac{3x-5}{1-x} \geqslant x-3$.

Solution

$$\frac{3x-5}{1-x} \geqslant x-3$$

$$\Rightarrow \quad (3x-5)(1-x) \geqslant (x-3)(1-x)^2$$

$$\Rightarrow \quad (3x-5)(1-x)-(x-3)(1-x)^2 \geqslant 0$$

$$\Rightarrow \quad (1-x)[(3x-5)-(x-3)(1-x)] \geqslant 0$$

$$\Rightarrow \quad (1-x)(3x-5+x^2-4x+3) \geqslant 0$$

$$\Rightarrow \quad (1-x)(x^2-x-2) \geqslant 0$$

$$\Rightarrow \quad (1-x)(x-2)(x+1) \geqslant 0$$

Let $f(x) = (1-x)(x-2)(x+1)$.

The critical values are $x=-1$, $x=1$, and $x=2$.

A sign diagram for $f(x)$ is given below:

$$f(-2) = (1 - -2)(-2-2)(-2+1) = (3)(-4)(-1) = 12$$
$$f(0) = (1-0)(0-2)(0+1) = (1)(-2)(1) = -2$$

	−1		1		2	
+		−		+		−

The question asks when the expression is greater than or equal to zero. There are two regions which gave a positive value for $f(x)$, so the solution is $x \leqslant -1$ or $1 < x \leqslant 2$.

Notice that the last inequality must have $1 < x$ since x cannot equal 1 in the original inequality.

EXERCISE 3A

Solve each of the following by multiplying throughout by the square of the denominator.

1 $\frac{x-3}{2-x} < 1$

2 $\frac{x-1}{x-4} > 2$

3 $\frac{x+5}{x-2} \leqslant 3$

4 $\frac{1}{x-7} < -2$

5 $\frac{6-x}{x+4} > 5$

6 $\frac{5}{x-6} \geqslant 4$

7 $\frac{2x-3}{x+4} < \frac{1}{3}$

8 $\frac{x}{x-3} \leqslant x$

9 $\frac{x(x-2)}{2x-5} \geqslant 3$

10 $\frac{x(x+5)}{x-4} > -2$

3.4 Combining terms into a single fraction

Another method for solving inequalities of this type is to take all terms onto one side of the inequality and combine them into a single algebraic fraction.

Worked example 3.4

Find the solution of the inequality $\frac{4x+5}{x-3} > 2$.

Solution

$$\frac{4x+5}{x-3} > 2 \Rightarrow \frac{4x+5}{x-3} - 2 > 0$$

You must have 0 on the right-hand side.

$$\Rightarrow \frac{4x+5}{x-3} - \frac{2(x-3)}{x-3} > 0$$

Subtract the fractions using a common denominator.

$$\Rightarrow \frac{4x+5-2x+6}{x-3} > 0$$

Take **great care** with the signs.

$$\Rightarrow \frac{(2x+11)}{(x-3)} > 0$$

Simplify. The brackets are not strictly necessary here but are used to emphasise the method.

Again, you find the critical values.

Let $f(x) = \frac{(2x+11)}{(x-3)}$.

The critical values are $x = -\frac{11}{2}$ and $x = 3$.

These values make each bracket in turn equal to 0.

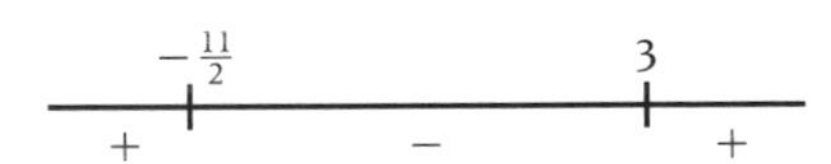

The critical values cut the line into three regions: $x < -\frac{11}{2}$, $-\frac{11}{2} < x < 3$ and $x > 3$.

The sign diagram is shown above:

$$f(-6) = \tfrac{1}{9}, \quad f(0) = -\tfrac{11}{3}, \quad f(4) = 19$$

You need to find when $\frac{(2x+11)}{(x-3)} > 0$.

There are two regions which gave a positive value for $f(x)$.

The solution is $x < -\frac{11}{2}$ or $x > 3$.

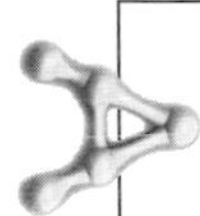

When solving inequalities involving a rational expression, you must ensure that the expression is written in an appropriate form (i.e. factorised/simplified as far as possible and with 0 on the right-hand side).

You can then find the critical values for the numerator and denominator.

Worked example 3.5

Solve the inequality $\frac{x+5}{x-3} < \frac{x-2}{x+4}$.

Solution

$$\frac{x+5}{x-3} < \frac{x-2}{x+4}$$

$$\Rightarrow \quad \frac{x+5}{x-3} - \frac{x-2}{x+4} < 0$$

$$\Rightarrow \quad \frac{(x+5)(x+4)}{(x-3)(x+4)} - \frac{(x-2)(x-3)}{(x-3)(x+4)} < 0$$

$$\Rightarrow \quad \frac{x^2+9x+20}{(x-3)(x+4)} - \frac{x^2-5x+6}{(x-3)(x+4)} < 0$$

$$\Rightarrow \quad \frac{x^2+9x+20-x^2+5x-6}{(x-3)(x+4)} < 0$$

$$\Rightarrow \quad \frac{(14x+14)}{(x-3)(x+4)} < 0$$

$$\Rightarrow \quad \frac{14(x+1)}{(x-3)(x+4)} < 0$$

Let $\mathrm{f}(x) = \frac{14(x+1)}{(x-3)(x+4)}$.

The critical values are $x = -4$, $x = -1$ and $x = 3$. A sign diagram can be drawn:

	-4		-1		3	
$-$		$+$		$-$		$+$

The critical values cut the line into four regions:
$x < -4$, $-4 < x < -1$, $-1 < x < 3$ and $x > 3$.

You need to find when $\frac{14(x+1)}{(x-3)(x+4)} < 0$.

There are two regions which give a negative value for $\mathrm{f}(x)$ so the solution is $x < -4$ or $-1 < x < 3$.

Finding a few values:

$$f(-5) = \frac{14(-5+1)}{(-5-3)(-5+4)} = \frac{-56}{(-8)(-1)} = -7$$

$$f(0) = \frac{14(0+1)}{(0-3)(0+4)} = \frac{14}{(-3)(4)} = -\frac{7}{6}$$

Worked example 3.6

Solve the inequality $\frac{2x+3}{x-6} \geqslant x-2$.

Note that the inequality is not defined when $x = 6$.

Solution

$$\frac{2x+3}{x-6} \geqslant x-2$$

$$\Rightarrow \quad \frac{2x+3}{x-6} - (x-2) \geqslant 0$$

$$\Rightarrow \quad \frac{2x+3}{x-6} - \frac{(x-2)(x-6)}{x-6} \geqslant 0$$

$$\Rightarrow \quad \frac{2x+3-(x^2-8x+12)}{x-6} \geqslant 0$$

$$\Rightarrow \quad \frac{-x^2+10x-9}{(x-6)} \geqslant 0$$

$$\Rightarrow \quad \frac{-(x^2-10x+9)}{(x-6)} \geqslant 0$$

$$\Rightarrow \quad \frac{-(x-1)(x-9)}{(x-6)} \geqslant 0$$

Let $f(x) = \frac{-(x-1)(x-9)}{(x-6)}$.

$$f(0) = \frac{-(0-1)(0-9)}{(0-6)} = \frac{-9}{-6} = \frac{3}{2}$$

$$f(7) = \frac{-(7-1)(7-9)}{(7-6)} = \frac{12}{1} = 12$$

The critical values are $x = 1$, $x = 6$ and $x = 9$. A sign diagram is drawn below:

	1		6		9	
+		−		+		−

A **very important point** here is that you cannot have $x = 6$ as part of a solution since the denominator of $f(x)$ becomes zero when $x = 6$. This means that the $\leqslant$ sign cannot be attached to the critical value 6.

The critical values suggest four possible intervals to be considered for the solution:

$x \leqslant 1$, $1 \leqslant x < 6$, $6 < x \leqslant 9$ and $x \geqslant 9$.

You need to find when $\frac{-(x-1)(x-9)}{(x-6)} \geqslant 0$.

There are two regions which give a positive value for $f(x)$.

The solution is therefore $x \leqslant 1$ or $6 < x \leqslant 9$.

EXERCISE 3B

Solve the following inequalities by taking all the terms onto one side and combining them into a single fraction:

1 $\dfrac{3x+1}{2x-5} > 1$

2 $\dfrac{7x-2}{x-3} < 5$

3 $\dfrac{5x-2}{x+2} \leqslant x$

4 $\dfrac{4}{x-2} < x+1$

5 $\dfrac{3}{x-1} > \dfrac{2}{x+2}$

6 $\dfrac{1}{x+4} > \dfrac{x}{x-2}$

7 $\dfrac{x}{x+3} > \dfrac{4}{x}$

8 $\dfrac{x}{x-3} \leqslant \dfrac{x}{x+2}$

9 $\dfrac{x+3}{x-1} \geqslant \dfrac{x-1}{x+4}$

10 $\dfrac{x-1}{x-3} < \dfrac{x+2}{x-5}$

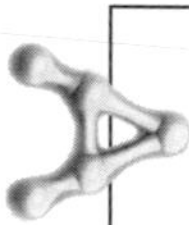

There are various methods for solving inequalities involving rational expressions, such as:

1 multiplying throughout by the square of the denominator;

2 combining all the fractions into one single term on one side of the inequality.

MIXED EXERCISE

Solve inequalities 1–10 by any appropriate method.

1 $\dfrac{5x-4}{x+2} \leqslant 3$

2 $\dfrac{x}{x+1} \leqslant \dfrac{x}{x+3}$

3 $\dfrac{6x+5}{x-1} \geqslant 4$

4 $\dfrac{3x-4}{2x} < 1$

5 $\dfrac{6}{x+5} > x$

6 $\dfrac{5x}{(x-3)^2} \geqslant 20$

7 $\dfrac{x+6}{x(x-2)} < 3$

8 $\dfrac{x-3}{x+5} > \dfrac{x-5}{x+2}$

9 $\dfrac{(x-2)(x+3)(x-4)}{(x-1)} < 0$

10 $\dfrac{2}{x+5} > x+4$

11 Solve the inequality $\dfrac{4x-3}{x-1} < x+3$. [A]

12 Solve the inequality $\dfrac{3x-5}{x-2} \leqslant 4$. [A]

13 A student attempts to solve the inequality $\frac{2x+3}{x-2} \geqslant 9$, and writes the following statements:

Step 1 $2x + 3 \geqslant 9(x - 2)$

Step 2 $2x + 3 \geqslant 9x - 18$

Step 3 $21 \geqslant 7x$

Step 4 $x \leqslant 3$

(a) Show that, although $x = 1$ satisfies the student's solution, it does not satisfy the original inequality.

(b) State, with a reason, the step where the student makes an error.

(c) Determine the correct solution to the inequality

$$\frac{2x+3}{x-2} \geqslant 9.$$

[A]

Key point summary

1 Important differences between solving equations and solving inequalities are: *p30*

1 An inequality will have a **range** of values as its solution.

2 Whenever you multiply or divide an inequality by a **negative** number you must also **reverse** the inequality sign.

2 When solving quadratic/cubic/higher order inequalities you need to consider the **critical values**. *p31*

You calculate $f(x)$ in each of the regions of the number line created by the critical values and produce a sign diagram.

3 When solving inequalities involving a rational expression, you must ensure that the expression is written in an appropriate form (i.e. factorised/simplified as far as possible and with 0 on the right-hand side). *p33*

You can then find the critical values for the numerator and denominator.

4 There are various methods for solving inequalities involving rational expressions, such as: *p36*

1 multiplying throughout by the square of the denominator;

2 combining all the fractions into one single term on one side of the inequality.

Test yourself	What to review
1 Solve the inequality $\dfrac{(2x+1)(x-2)}{(x+5)} > 0$	*Sections 3.3, 3.4*
2 Solve the inequality $\dfrac{4x+5}{x-2} < 3$.	*Sections 3.3, 3.4*
3 Solve the inequality $\dfrac{x+5}{x-7} \leqslant x-2$.	*Sections 3.3, 3.4*
4 Solve $\dfrac{x-3}{x+6} > \dfrac{x-2}{x+2}$	*Sections 3.3, 3.4*

Test yourself ANSWERS

1 $-5 < x < -\frac{1}{2}$ or $x > 2$

2 $-11 < x < 2$

3 $1 \leqslant x < 7$ or $x \geqslant 9$

4 $x < -6$ or $-2 < x < \frac{6}{5}$

CHAPTER 4

Matrices

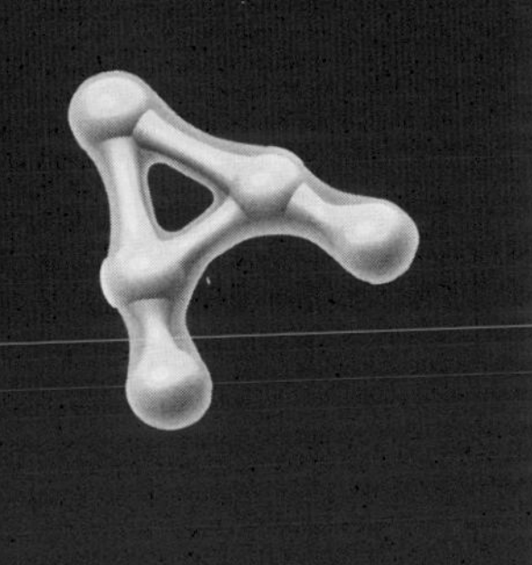

Learning objectives

After studying this chapter, you should be able to:

- understand what matrices are
- add, subtract and multiply compatible matrices.

4.1 Introduction

A **matrix** is a **store for information**.

For example, the number of boys and girls in Years 12 and 13 of a college could be recorded as a matrix:

$$\begin{array}{l} \\ \text{Year 12} \\ \text{Year 13} \end{array} \begin{array}{c} \text{boys} \quad \text{girls} \\ \begin{bmatrix} 65 & 67 \\ 68 & 61 \end{bmatrix} \end{array} \qquad \text{or simply } \mathbf{A} = \begin{bmatrix} 65 & 67 \\ 68 & 61 \end{bmatrix}$$

A bold capital letter is often used to signify a matrix.

So matrices are essentially stores for **numbers**.

Matrices is the plural of matrix.

They are rectangular in appearance and are made up of a number of entries or **elements**.

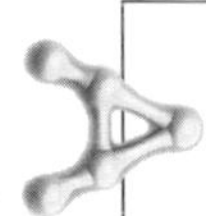

A **matrix** is a rectangular array of numbers. Each entry in the matrix is called an **element**.

4.2 The order of a matrix

Matrices can come in different sizes. Consider the following:

$$\mathbf{A} = \begin{bmatrix} 4 & 2 \\ -2 & 6 \end{bmatrix} \quad \mathbf{B} = \begin{bmatrix} 7 & 3 & 0 \\ -2 & 1 & 9 \\ 7 & 2 & 1 \end{bmatrix} \quad \mathbf{C} = \begin{bmatrix} 4 & 9 & 1 \\ -3 & 0 & 1 \end{bmatrix}$$

Matrices are classified by the number of rows and the number of columns that they have.

In the above:
matrix **A** has 2 rows and 2 columns, it is a 2×2 matrix;
matrix **B** has 3 rows and 3 columns, it is a 3×3 matrix;
matrix **C** has 2 rows and 3 columns, it is a 2×3 matrix.

Read as '2 by 2'.

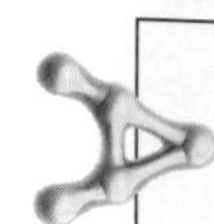

A matrix with m rows and n columns is an $m \times n$ matrix. This is called the **order** of the matrix.

In this module you will consider matrices of order 2×2 or 2×1 and in further modules matrices of higher orders will be used.

Note: You may see matrices written using round brackets instead of square brackets, e.g. you could see either

$$\mathbf{A} = \begin{pmatrix} 2 & 5 \\ 3 & -1 \end{pmatrix} \quad \text{or} \quad \mathbf{A} = \begin{bmatrix} 2 & 5 \\ 3 & -1 \end{bmatrix}.$$

Both are perfectly acceptable.

4.3 Adding and subtracting matrices

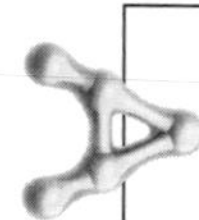

We can add or subtract matrices provided they have the **same order**.

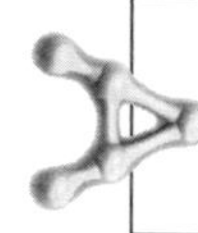

Adding/subtracting matrices is a very simple process since all we do is add/subtract **corresponding elements** from each matrix.

Worked example 4.1

The matrices **A**, **B**, **C** and **D** are given by

$$\mathbf{A} = \begin{bmatrix} 3 & 6 \\ 1 & 3 \end{bmatrix}, \quad \mathbf{B} = \begin{bmatrix} 7 & -2 \\ -1 & 4 \end{bmatrix}, \quad \mathbf{C} = \begin{bmatrix} 9 \\ -4 \end{bmatrix} \quad \text{and} \quad \mathbf{D} = \begin{bmatrix} 5 \\ -2 \end{bmatrix}.$$

Find the following where possible:

(a) $\mathbf{A} + \mathbf{B}$, **(b)** $\mathbf{A} - \mathbf{B}$, **(c)** $\mathbf{C} + \mathbf{D}$,

(d) $\mathbf{D} - \mathbf{C}$, **(e)** $\mathbf{A} + \mathbf{C}$.

Solution

(a) $$\mathbf{A} + \mathbf{B} = \begin{bmatrix} 3 & 6 \\ 1 & 3 \end{bmatrix} + \begin{bmatrix} 7 & -2 \\ -1 & 4 \end{bmatrix}$$

$$= \begin{bmatrix} 3 + 7 & 6 + -2 \\ 1 + -1 & 3 + 4 \end{bmatrix} = \begin{bmatrix} 10 & 4 \\ 0 & 7 \end{bmatrix}$$

(b) $$\mathbf{A} - \mathbf{B} = \begin{bmatrix} 3 & 6 \\ 1 & 3 \end{bmatrix} - \begin{bmatrix} 7 & -2 \\ -1 & 4 \end{bmatrix}$$

$$= \begin{bmatrix} 3 - 7 & 6 - -2 \\ 1 - -1 & 3 - 4 \end{bmatrix} = \begin{bmatrix} -4 & 8 \\ 2 & -1 \end{bmatrix}$$

(c) $\mathbf{C} + \mathbf{D} = \begin{bmatrix} 9 \\ -4 \end{bmatrix} + \begin{bmatrix} 5 \\ -2 \end{bmatrix} = \begin{bmatrix} 9+5 \\ -4+-2 \end{bmatrix} = \begin{bmatrix} 14 \\ -6 \end{bmatrix}$

(d) $\mathbf{D} - \mathbf{C} = \begin{bmatrix} 5 \\ -2 \end{bmatrix} - \begin{bmatrix} 9 \\ -4 \end{bmatrix} = \begin{bmatrix} 5-\ \ 9 \\ -2-\ -4 \end{bmatrix} = \begin{bmatrix} -4 \\ 2 \end{bmatrix}$

(e) $\mathbf{A} + \mathbf{C} = \begin{bmatrix} 3 & 6 \\ 1 & 3 \end{bmatrix} + \begin{bmatrix} 9 \\ -4 \end{bmatrix}$

This is impossible to calculate since the matrices have different orders.

EXERCISE 4A

1 Write down the order of each of the following matrices.

(a) $\begin{bmatrix} 5 & 4 & -9 & 2 \\ 7 & 4 & 0 & 0 \end{bmatrix}$ **(b)** $\begin{bmatrix} 3 & 9 \\ 9 & -2 \end{bmatrix}$

(c) $[6\ \ 5\ \ 0]$ **(d)** $\begin{bmatrix} 4 & 0 \\ 1 & -3 \\ 4 & -2 \end{bmatrix}$

2 Find:

(a) $\begin{bmatrix} 4 \\ 5 \end{bmatrix} + \begin{bmatrix} -4 \\ 10 \end{bmatrix}$

(b) $\begin{bmatrix} 20 \\ -4 \end{bmatrix} + \begin{bmatrix} -15 \\ -3 \end{bmatrix}$

(c) $\begin{bmatrix} 7 & 4 \\ -3 & 2 \end{bmatrix} + \begin{bmatrix} 2 & 11 \\ -5 & 1 \end{bmatrix}$

(d) $\begin{bmatrix} 6 & 3 \\ -1 & 6 \end{bmatrix} + \begin{bmatrix} -2 & 1 \\ 5 & 5 \end{bmatrix}$

(e) $\begin{bmatrix} 2 & 10 \\ 9 & -3 \end{bmatrix} + \begin{bmatrix} -3 & 6 \\ 8 & -1 \end{bmatrix} + \begin{bmatrix} 5 & 7 \\ -3 & 7 \end{bmatrix}$

3 Find:

(a) $\begin{bmatrix} 6 \\ 7 \end{bmatrix} - \begin{bmatrix} 4 \\ -3 \end{bmatrix}$

(b) $\begin{bmatrix} 5 \\ -8 \end{bmatrix} - \begin{bmatrix} 3 \\ -2 \end{bmatrix}$

(c) $\begin{bmatrix} 8 & 2 \\ -2 & 3 \end{bmatrix} - \begin{bmatrix} 3 & 9 \\ 6 & -1 \end{bmatrix}$

(d) $\begin{bmatrix} 7 & 10 \\ 15 & -3 \end{bmatrix} - \begin{bmatrix} 9 & 10 \\ 5 & 7 \end{bmatrix}$

(e) $\begin{bmatrix} 5 & -2 \\ 1 & -2 \end{bmatrix} + \begin{bmatrix} 1 & 9 \\ -3 & 7 \end{bmatrix} - \begin{bmatrix} 6 & -6 \\ -1 & 3 \end{bmatrix}$

4 The matrices **A**, **B**, and **C** are given by:

$$\mathbf{A} = \begin{bmatrix} 6 & -3 \\ 9 & 17 \end{bmatrix}, \quad \mathbf{B} = \begin{bmatrix} -3 & 9 \\ 0 & -4 \end{bmatrix} \text{ and } \mathbf{C} = \begin{bmatrix} 10 & -5 \\ 1 & 4 \end{bmatrix}.$$

Find:

(a) $\mathbf{A} + \mathbf{B}$, **(b)** $\mathbf{B} + \mathbf{A}$, **(c)** $\mathbf{C} - \mathbf{A}$,
(d) $\mathbf{B} + \mathbf{C}$, **(e)** $\mathbf{A} - \mathbf{B}$, **(f)** $\mathbf{A} + \mathbf{B} + \mathbf{C}$,
(g) $\mathbf{B} + \mathbf{C} - \mathbf{A}$, **(h)** $\mathbf{A} - \mathbf{C} + \mathbf{B}$.

4.4 Multiples of matrices

Suppose $\mathbf{M} = \begin{bmatrix} 3 & -2 \\ 0 & 4 \end{bmatrix}$, then $\mathbf{M} + \mathbf{M} + \mathbf{M} = \begin{bmatrix} 9 & -6 \\ 0 & 12 \end{bmatrix}$.

This is equivalent to 3**M** and each element of matrix **M** has been multiplied by 3.

It is therefore a simple matter to multiply a matrix by a constant.

For example, $5\begin{bmatrix} 4 & 1 & 0 \\ -2 & 5 & -9 \end{bmatrix} = \begin{bmatrix} 20 & 5 & 0 \\ -10 & 25 & -45 \end{bmatrix}$.

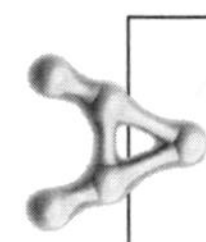

To multiply a matrix by a constant, simply multiply each element of the matrix by the constant.

Worked example 4.2

Given that $\mathbf{A} = \begin{bmatrix} 3 & 6 \\ 4 & -5 \end{bmatrix}$ and $\mathbf{B} = \begin{bmatrix} -2 & 0 \\ 5 & 4 \end{bmatrix}$, find:

(a) 3**A**, **(b)** 7**B**, **(c)** 4**A** + 3**B**, **(d)** 5**B** − 2**A**.

Solution

(a) $3\mathbf{A} = 3\begin{bmatrix} 3 & 6 \\ 4 & -5 \end{bmatrix} = \begin{bmatrix} 9 & 18 \\ 12 & -15 \end{bmatrix}$

(b) $7\mathbf{B} = 7\begin{bmatrix} -2 & 0 \\ 5 & 4 \end{bmatrix} = \begin{bmatrix} -14 & 0 \\ 35 & 28 \end{bmatrix}$

(c) $4\mathbf{A} + 3\mathbf{B} = 4\begin{bmatrix} 3 & 6 \\ 4 & -5 \end{bmatrix} + 3\begin{bmatrix} -2 & 0 \\ 5 & 4 \end{bmatrix}$

$$= \begin{bmatrix} 12 & 24 \\ 16 & -20 \end{bmatrix} + \begin{bmatrix} -6 & 0 \\ 15 & 12 \end{bmatrix} = \begin{bmatrix} 6 & 24 \\ 31 & -8 \end{bmatrix}$$

(d) $5\mathbf{B} - 2\mathbf{A} = 5\begin{bmatrix} -2 & 0 \\ 5 & 4 \end{bmatrix} - 2\begin{bmatrix} 3 & 6 \\ 4 & -5 \end{bmatrix}$

$$= \begin{bmatrix} -10 & 0 \\ 25 & 20 \end{bmatrix} - \begin{bmatrix} 6 & 12 \\ 8 & -10 \end{bmatrix} = \begin{bmatrix} -16 & -12 \\ 17 & 30 \end{bmatrix}$$

4.5 Multiplying two matrices

You can multiply a row matrix by a column matrix provided they have the same number of elements. You multiply corresponding elements and add the products together.

Hence,

$$[2 \quad 5 \quad 7]\begin{bmatrix}3\\1\\4\end{bmatrix} = [2\times 3 \quad + \quad 5\times 1 \quad + \quad 7\times 4] = [6+5+28]$$
$$= [39]$$

When you multiply two matrices together you are essentially multiplying the rows of the first matrix with the columns of the second.

In order to be able to do this, the number of columns in the first matrix must be the same as the number of rows in the second.

We can multiply two matrices **A** and **B** only if the number of columns of **A** equals the number of rows of **B**.

It is rather like playing dominos, you can line up a domino next to a giving the situation .
The effect is as if you had the situation .

If **A** is a matrix of order $a \times b$ and **B** is a matrix of order $c \times d$, then the matrix **AB** exists if and only if $b = c$.

The product **AB** will have order $a \times d$.

The method for multiplying two compatible matrices is illustrated in the following examples.

Worked example 4.3

Find the product **AB** when $\mathbf{A} = \begin{bmatrix}4 & 9\\10 & 3\end{bmatrix}$ and $\mathbf{B} = \begin{bmatrix}5\\8\end{bmatrix}$.

Solution

Firstly,

	A	**B**
order	2×2	2×1

2×2 and 2×1: ↑ ↑ the same

Check for compatibility.

The number of columns of **A** equals the number of rows of **B**, therefore the product matrix exists.

The product matrix will have order 2×1.

The method for multiplying the two matrices is illustrated step-by-step below:

$$\mathbf{AB} = \begin{bmatrix} 4 & 9 \\ 10 & 3 \end{bmatrix} \begin{bmatrix} 5 \\ 8 \end{bmatrix} = \begin{bmatrix} ? \\ ? \end{bmatrix}$$

Find the element of the first row:

$$= \begin{bmatrix} (4 \times 5) + (9 \times 8) \\ ? \end{bmatrix} = \begin{bmatrix} 92 \\ ? \end{bmatrix}$$

First row of **A** multiplied by first column of **B**.

Find the first element of the second row:

$$= \begin{bmatrix} 92 \\ (10 \times 5) + (3 \times 8) \end{bmatrix} = \begin{bmatrix} 92 \\ 74 \end{bmatrix}$$

Second row of **A** multiplied by the first column of **B**.

So the product $\mathbf{AB} = \begin{bmatrix} 92 \\ 74 \end{bmatrix}$.

The same process can be applied when multiplying a 2×2 matrix with another 2×2 matrix.

Worked example 4.4

Find the product **CD** when $\mathbf{C} = \begin{bmatrix} 2 & 3 \\ 5 & 8 \end{bmatrix}$ and $\mathbf{D} = \begin{bmatrix} 4 & 9 \\ 6 & 7 \end{bmatrix}$.

Solution

	C	**D**
order	2×2	2×2
	↑	↑
	the same	

Check for compatibility.

The number of columns of **C** equals the number of rows of **D**, therefore the product matrix exists.
The product matrix will have order 2×2.

The method for multiplying the two matrices is illustrated step-by-step below:

$$\mathbf{CD} = \begin{bmatrix} 2 & 3 \\ 5 & 8 \end{bmatrix} \begin{bmatrix} 4 & 9 \\ 6 & 7 \end{bmatrix} = \begin{bmatrix} ? & ? \\ ? & ? \end{bmatrix}$$

Find the first element of the first row:

$$= \begin{bmatrix} (2 \times 4) + (3 \times 6) & ? \\ ? & ? \end{bmatrix} = \begin{bmatrix} 26 & ? \\ ? & ? \end{bmatrix}$$

First row of **C** multiplied by first column of **D**.

Find the second element of the first row:

$$= \begin{bmatrix} 26 & (2 \times 9) + (3 \times 7) \\ ? & ? \end{bmatrix} = \begin{bmatrix} 26 & 39 \\ ? & ? \end{bmatrix}$$

First row of **C** multiplied by the second column of **D**.

Find the first element of the second row:

$$= \begin{bmatrix} 26 & 39 \\ (5 \times 4) + (8 \times 6) & ? \end{bmatrix} = \begin{bmatrix} 26 & 39 \\ 68 & ? \end{bmatrix}$$

Second row of **C** multiplied by the first column of **D**.

Find the second element of the second row:

$$= \begin{bmatrix} 26 & 39 \\ 68 & (5 \times 9) + (8 \times 7) \end{bmatrix} = \begin{bmatrix} 26 & 39 \\ 68 & 101 \end{bmatrix}$$

Second row of **C** multiplied by the second column of **D**.

So the product $\mathbf{CD} = \begin{bmatrix} 26 & 39 \\ 68 & 101 \end{bmatrix}$.

Worked example 4.5

Find **PQ** if $\mathbf{P} = \begin{bmatrix} 3 & 5 \\ 0 & 1 \end{bmatrix}$ and $\mathbf{Q} = \begin{bmatrix} 4 & 2 \\ -6 & 1 \end{bmatrix}$.

Solution

Firstly, the matrices are of order: $\underset{2 \times 2}{\mathbf{P}} \quad \underset{2 \times 2}{\mathbf{Q}}$.

Check for compatibility.

The matrices are therefore compatible and the product matrix has order 2×2.

$\mathbf{2 \times 2 \quad 2 \times 2}$

$$\mathbf{PQ} = \begin{bmatrix} 3 & 5 \\ 0 & 1 \end{bmatrix}\begin{bmatrix} 4 & 2 \\ -6 & 1 \end{bmatrix}$$

$$= \begin{bmatrix} (3 \times 4) + (5 \times -6) & (3 \times 2) + (5 \times 1) \\ (0 \times 4) + (1 \times -6) & (0 \times 2) + (1 \times 1) \end{bmatrix} = \begin{bmatrix} -18 & 11 \\ -6 & 1 \end{bmatrix}$$

Worked example 4.6

Find the value of the constant k for which $\begin{bmatrix} 7 & -1 \\ 4 & 2 \end{bmatrix}\begin{bmatrix} -1 \\ -1 \end{bmatrix} = k\begin{bmatrix} 1 \\ 1 \end{bmatrix}$.

Solution

$$\begin{bmatrix} 7 & -1 \\ 4 & 2 \end{bmatrix}\begin{bmatrix} -1 \\ -1 \end{bmatrix} = \begin{bmatrix} 7(-1) + -1(-1) \\ 4(-1) + 2(-1) \end{bmatrix} = \begin{bmatrix} -6 \\ -6 \end{bmatrix} = -6\begin{bmatrix} 1 \\ 1 \end{bmatrix}$$

$\Rightarrow \quad k = -6$

Worked example 4.7

Find a matrix **M** such that $\mathbf{M}\begin{bmatrix} 5 & -2 \\ -3 & 4 \end{bmatrix} = \begin{bmatrix} 23 & -12 \\ 9 & 2 \end{bmatrix}$.

Solution

Let matrix $\mathbf{M} = \begin{bmatrix} a & b \\ c & d \end{bmatrix}$

M must be a 2×2 to give the correct order for the product matrix.

So $\mathbf{M}\begin{bmatrix} 5 & -2 \\ -3 & 4 \end{bmatrix} = \begin{bmatrix} a & b \\ c & d \end{bmatrix}\begin{bmatrix} 5 & -2 \\ -3 & 4 \end{bmatrix} = \begin{bmatrix} 5a - 3b & -2a + 4b \\ 5c - 3d & -2c + 4d \end{bmatrix}$

$\Rightarrow \quad \begin{bmatrix} 5a - 3b & -2a + 4b \\ 5c - 3d & -2c + 4d \end{bmatrix} = \begin{bmatrix} 23 & -12 \\ 9 & 2 \end{bmatrix}$

If these two matrices are the same then **corresponding elements must be the same**.

We can use this fact to generate two pairs of simultaneous equations: one pair involving a and b and the other pair involving c and d.

$$5a - 3b = 23 \Rightarrow 20a - 12b = 92 \quad [1]$$
$$-2a + 4b = -12 \Rightarrow -6a + 12b = -36 \quad [2]$$

[1] + [2] gives $14a = 56 \Rightarrow a = 4$

$\Rightarrow b = -1$

Also,

$$5c - 3d = 9 \Rightarrow 20c - 12d = 36 \quad [3]$$
$$-2c + 4d = 2 \Rightarrow -6c + 12d = 6 \quad [4]$$

[3] + [4] gives $14c = 42 \Rightarrow c = 3$

$\Rightarrow d = 2$

So the matrix $\mathbf{M}$ is given by $\mathbf{M} = \begin{bmatrix} 4 & -1 \\ 3 & 2 \end{bmatrix}$.

Worked example 4.8

Given that $\mathbf{A} = \begin{bmatrix} 4 & 2 \\ -3 & 1 \end{bmatrix}$ and $\mathbf{B} = \begin{bmatrix} 6 & 0 \\ -2 & 3 \end{bmatrix}$ find:

(a) **AB**, **(b)** **BA**.

Solution

(a) $\mathbf{AB} = \begin{bmatrix} 4 & 2 \\ -3 & 1 \end{bmatrix}\begin{bmatrix} 6 & 0 \\ -2 & 3 \end{bmatrix} = \begin{bmatrix} 20 & 6 \\ -20 & 3 \end{bmatrix}$

(b) $\mathbf{BA} = \begin{bmatrix} 6 & 0 \\ -2 & 3 \end{bmatrix}\begin{bmatrix} 4 & 2 \\ -3 & 1 \end{bmatrix} = \begin{bmatrix} 24 & 12 \\ -17 & -1 \end{bmatrix}$

It should be clear from the examples above that the order in which we multiply two matrices is very important since it makes a difference to the answer.

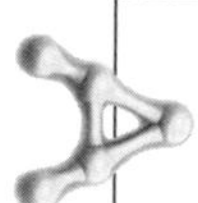

In general, $\mathbf{AB} \neq \mathbf{BA}$ (usually).

Matrix multiplication is not, in general, commutative.

In **AB**, we say that **B** is **pre-multiplied** by **A**.
In **BA**, we say that **B** is **post-multiplied** by **A**.

4.6 Special matrices

The next few sections deal with matrices that have a certain significance or importance.

Square matrices

A **square matrix** is one which has the same number of rows as columns.
A square matrix has order $m \times m$.

A matrix which has the same number of rows and columns is called a **square matrix**.

For example, $\begin{bmatrix} 3 & 5 \\ 1 & -2 \end{bmatrix}$, $\begin{bmatrix} 3 & 6 & 0 \\ 1 & -2 & -6 \\ 9 & 2 & 5 \end{bmatrix}$ and $\begin{bmatrix} 0 & 2 & 7 & -3 \\ 9 & 5 & 1 & -2 \\ -3 & 4 & 8 & 2 \\ 9 & -3 & 2 & 7 \end{bmatrix}$

are all square matrices of order 2×2, 3×3 and 4×4, respectively.

You will meet 2×2 matrices again in a later chapter when dealing with matrix transformations.

The identity matrix

Consider the 2×2 matrix $\mathbf{A} = \begin{bmatrix} a & b \\ c & d \end{bmatrix}$.

Notice what happens when we multiply this matrix by $\begin{bmatrix} 1 & 0 \\ 0 & 1 \end{bmatrix}$.

$$\begin{bmatrix} a & b \\ c & d \end{bmatrix}\begin{bmatrix} 1 & 0 \\ 0 & 1 \end{bmatrix} = \begin{bmatrix} a & b \\ c & d \end{bmatrix}$$

also $$\begin{bmatrix} 1 & 0 \\ 0 & 1 \end{bmatrix}\begin{bmatrix} a & b \\ c & d \end{bmatrix} = \begin{bmatrix} a & b \\ c & d \end{bmatrix}$$

You should, of course, notice that you get the matrix **A** as the answer irrespective of the order of multiplication.

The matrix $\mathbf{I} = \begin{bmatrix} 1 & 0 \\ 0 & 1 \end{bmatrix}$ is called the 2×2 **identity matrix** because when you multiply any 2×2 matrix **A** by **I** you get **A** as the answer.
This means that for any 2×2 matrix **A**,

$$\mathbf{IA} = \mathbf{AI} = \mathbf{A}.$$

In other words, multiplication by **I** (either pre-multiplication or post-multiplication) leaves the elements of **A** unchanged.

The **identity matrix** in matrix multiplication is rather like the number 1 in 'ordinary' multiplication. (Multiplying any number by 1 leaves a number unchanged.)

We can actually have identity matrices of any size (as long as they are square matrices).

E.g. the matrix $\begin{bmatrix} 1 & 0 & 0 \\ 0 & 1 & 0 \\ 0 & 0 & 1 \end{bmatrix}$ is called the 3×3 identity matrix.

The zero matrix

Consider the matrix $\mathbf{A} = \begin{bmatrix} a & b \\ c & d \end{bmatrix}$.

Notice what happens when we add the matrix $\begin{bmatrix} 0 & 0 \\ 0 & 0 \end{bmatrix}$.

$$\begin{bmatrix} a & b \\ c & d \end{bmatrix} + \begin{bmatrix} 0 & 0 \\ 0 & 0 \end{bmatrix} = \begin{bmatrix} a & b \\ c & d \end{bmatrix}$$

also $\begin{bmatrix} 0 & 0 \\ 0 & 0 \end{bmatrix} + \begin{bmatrix} a & b \\ c & d \end{bmatrix} = \begin{bmatrix} a & b \\ c & d \end{bmatrix}$

Similarly, multiplying by this matrix gives

$$\begin{bmatrix} a & b \\ c & d \end{bmatrix}\begin{bmatrix} 0 & 0 \\ 0 & 0 \end{bmatrix} = \begin{bmatrix} 0 & 0 \\ 0 & 0 \end{bmatrix}$$

also $\begin{bmatrix} 0 & 0 \\ 0 & 0 \end{bmatrix}\begin{bmatrix} a & b \\ c & d \end{bmatrix} = \begin{bmatrix} 0 & 0 \\ 0 & 0 \end{bmatrix}$

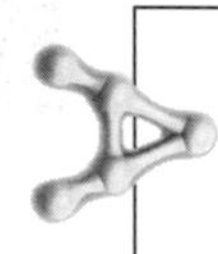

The 2×2 matrix $\mathbf{Z} = \begin{bmatrix} 0 & 0 \\ 0 & 0 \end{bmatrix}$ is called the **zero matrix**

since $\mathbf{Z} + \mathbf{A} = \mathbf{A} + \mathbf{Z} = \mathbf{A}$

and $\mathbf{ZA} = \mathbf{AZ} = \mathbf{Z}$ for any 2×2 matrix $\mathbf{A}$.

So **Z** leaves any matrix **A** unchanged under matrix addition, and itself remains unchanged under matrix multiplication. The matrix **Z** corresponds to the number 0 in 'ordinary' arithmetic.

Worked example 4.9

Find **AB**, where $\mathbf{A} = \begin{bmatrix} 2 & 1 \\ 10 & 6 \end{bmatrix}$ and $\mathbf{B} = \begin{bmatrix} 3 & -\frac{1}{2} \\ -5 & 1 \end{bmatrix}$.

What do you notice?

Solution

$$\mathbf{AB} = \begin{bmatrix} 2 & 1 \\ 10 & 6 \end{bmatrix}\begin{bmatrix} 3 & -\frac{1}{2} \\ -5 & 1 \end{bmatrix} = \begin{bmatrix} 1 & 0 \\ 0 & 1 \end{bmatrix}$$

You notice that the product matrix **AB** is the identity matrix.

Worked example 4.10

Matrix **A** is given by $\mathbf{A} = \begin{bmatrix} 3 & -1 \\ -10 & 4 \end{bmatrix}$.

Find the matrix **B** such that $\mathbf{AB} = \mathbf{I}$.

Solution

Firstly, for the product matrix to be 2×2, matrix **B** must also be 2×2.

Let $\mathbf{B} = \begin{bmatrix} a & b \\ c & d \end{bmatrix}$ and you need to find a, b, c and d.

Now $\mathbf{AB} = \mathbf{I}$

$$\Rightarrow \begin{bmatrix} 3 & -1 \\ -10 & 4 \end{bmatrix}\begin{bmatrix} a & b \\ c & d \end{bmatrix} = \begin{bmatrix} 1 & 0 \\ 0 & 1 \end{bmatrix}$$

Multiplying out the left-hand side gives:

$$\begin{bmatrix} 3a - c & 3b - d \\ -10a + 4c & -10b + 4d \end{bmatrix} = \begin{bmatrix} 1 & 0 \\ 0 & 1 \end{bmatrix}.$$

Equating terms in the first column of each side gives:

$$3a - c = 1$$
$$-10a + 4c = 0$$

Solving these equations simultaneously gives $a = 2$ and $c = 5$.

Equating terms in the second column of each side gives:

$$3b - d = 0$$
$$-10b + 4d = 1$$

Solving these equations simultaneously gives $b = \frac{1}{2}$ and $d = \frac{3}{2}$.

So matrix $\mathbf{B} = \begin{bmatrix} 2 & \frac{1}{2} \\ 5 & 1\frac{1}{2} \end{bmatrix}$.

EXERCISE 4B

1 Find the following matrix products:

(a) $\begin{bmatrix} 6 & 2 \\ 4 & 1 \end{bmatrix}\begin{bmatrix} 3 \\ 7 \end{bmatrix}$ **(b)** $\begin{bmatrix} 9 & 0 \\ 4 & 7 \end{bmatrix}\begin{bmatrix} 5 \\ 10 \end{bmatrix}$

(c) $\begin{bmatrix} 3 & 7 \\ -4 & 2 \end{bmatrix}\begin{bmatrix} 5 \\ 6 \end{bmatrix}$ **(d)** $\begin{bmatrix} 5 & -9 \\ 2 & 3 \end{bmatrix}\begin{bmatrix} 4 \\ 3 \end{bmatrix}$

(e) $\begin{bmatrix} 11 & 4 \\ -3 & -10 \end{bmatrix}\begin{bmatrix} 4 \\ -2 \end{bmatrix}$ **(f)** $\begin{bmatrix} 4 & -8 \\ -4 & 1 \end{bmatrix}\begin{bmatrix} -3 \\ 12 \end{bmatrix}$

2 Find each of the following:

(a) $\begin{bmatrix} 4 & 7 \\ 11 & 2 \end{bmatrix}\begin{bmatrix} 7 & 1 \\ 2 & 2 \end{bmatrix}$ **(b)** $\begin{bmatrix} 3 & 9 \\ -2 & 3 \end{bmatrix}\begin{bmatrix} 1 & 0 \\ 0 & 6 \end{bmatrix}$

(c) $\begin{bmatrix} -5 & 2 \\ 8 & -1 \end{bmatrix}\begin{bmatrix} -2 & 6 \\ 4 & 2 \end{bmatrix}$ **(d)** $\begin{bmatrix} -2 & 10 \\ 0 & 5 \end{bmatrix}\begin{bmatrix} 7 & -1 \\ -5 & 3 \end{bmatrix}$

3 Given that matrix $\mathbf{H} = \begin{bmatrix} 3 & 1 \\ -2 & -4 \end{bmatrix}$ find: **(a)** $\mathbf{H}^2$, **(b)** $\mathbf{H}^3$.

4 Find the values of x and y in the following cases:

(a) $\begin{bmatrix} 4 & 1 \\ -3 & 2 \end{bmatrix}\begin{bmatrix} 0 \\ -5 \end{bmatrix} = \begin{bmatrix} x \\ y \end{bmatrix}$ **(b)** $\begin{bmatrix} -4 & 5 \\ 4 & 2 \end{bmatrix}\begin{bmatrix} -6 \\ 8 \end{bmatrix} = \begin{bmatrix} x \\ y \end{bmatrix}$

(c) $\begin{bmatrix} 3 & 6 \\ -2 & -1 \end{bmatrix}\begin{bmatrix} x \\ y \end{bmatrix} = \begin{bmatrix} 0 \\ -6 \end{bmatrix}$ **(d)** $\begin{bmatrix} -2 & 1 \\ -5 & 6 \end{bmatrix}\begin{bmatrix} x \\ y \end{bmatrix} = \begin{bmatrix} -3 \\ -4 \end{bmatrix}$

(e) $\begin{bmatrix} x & 0 \\ y & 3 \end{bmatrix}\begin{bmatrix} 4 \\ 1 \end{bmatrix} = \begin{bmatrix} 16 \\ -21 \end{bmatrix}$ **(f)** $\begin{bmatrix} 8 & x \\ y & 2 \end{bmatrix}\begin{bmatrix} -3 \\ 2 \end{bmatrix} = \begin{bmatrix} -28 \\ -5 \end{bmatrix}$

5 Find the value of the constant c for which

$$\begin{bmatrix} 3 & 5 \\ -2 & 10 \end{bmatrix}\begin{bmatrix} 2 \\ 2 \end{bmatrix} = c\begin{bmatrix} 1 \\ 1 \end{bmatrix}.$$

6 Find a matrix **B** for which $\mathbf{B}\begin{bmatrix} 3 & 4 \\ 2 & -2 \end{bmatrix} = \begin{bmatrix} 16 & 12 \\ 7 & -14 \end{bmatrix}$.

7 Find a matrix **C** such that $\begin{bmatrix} 3 & 1 \\ -4 & 5 \end{bmatrix}\mathbf{C} = \begin{bmatrix} 10 & -3 \\ 12 & 4 \end{bmatrix}$.

8 Find two 2×2 matrices **A** and **B** such that **AB** = **Z** without any of the elements in **A** or **B** being zero.

9 Given that matrix $\mathbf{A} = \begin{bmatrix} 6 & 2 \\ 8 & 3 \end{bmatrix}$, find the matrix **B** such that **AB** = **I**.

10 Given that matrix $\mathbf{C} = \begin{bmatrix} -2 & 4 \\ 1 & 3 \end{bmatrix}$, find the matrix **D** such that **CD** = **I**.

Key point summary

1 A **matrix** is a rectangular array of numbers. Each entry in the matrix is called an **element**.	p39
2 A matrix with m rows and n columns is an $m \times n$ matrix. This is called the **order** of the matrix.	p40
3 We can add or subtract matrices provided they have the **same order**.	p40
4 To add/subtract matrices you add/subtract **corresponding elements**.	p40
5 To multiply a matrix by a constant, simply multiply each element of the matrix by the constant.	p42
6 We can multiply two matrices **A** and **B** only if the number of columns of **A** equals the number of rows of **B**.	p43
7 If **A** is an $(a \times b)$ matrix and **B** is a $(c \times d)$ matrix then the product matrix **AB** exists if and only if $b = c$. The product **AB** will be of order $a \times d$.	p43
8 In general, **AB** ≠ **BA**. Matrix multiplication is not, in general, commutative.	p46
9 A matrix which has the same number of rows and columns is called a **square matrix**.	p47

10 The matrix $\mathbf{I} = \begin{bmatrix} 1 & 0 \\ 0 & 1 \end{bmatrix}$ is called the 2×2 **identity matrix** because when you multiply any 2×2 matrix **A** by **I** you get **A** as the answer. *p47*
This means that for any 2×2 matrix **A**,

$$\mathbf{IA} = \mathbf{AI} = \mathbf{A}.$$

11 The 2×2 matrix $\mathbf{Z} = \begin{bmatrix} 0 & 0 \\ 0 & 0 \end{bmatrix}$ is called the **zero matrix** *p48*

since $\mathbf{Z} + \mathbf{A} = \mathbf{A} + \mathbf{Z} = \mathbf{A}$

and $\mathbf{ZA} = \mathbf{AZ} = \mathbf{Z}$ for any 2×2 matrix **A**.

Test yourself

What to review

1 If matrices **A**, **B** and **C** are given by *Sections 4.3, 4.4, 4.5*

$$\mathbf{A} = \begin{bmatrix} 3 & 2 \\ -1 & 0 \end{bmatrix}, \quad \mathbf{B} = \begin{bmatrix} 6 & -3 \\ -2 & 4 \end{bmatrix} \quad \text{and} \quad \mathbf{C} = \begin{bmatrix} 5 & 0 \\ 1 & -3 \end{bmatrix}, \text{ evaluate:}$$

(a) $\mathbf{A} + 2\mathbf{C}$ **(b)** $\mathbf{A} + \mathbf{B} - \mathbf{C}$
(c) $3\mathbf{A} + 2\mathbf{B}$ **(d)** $\mathbf{AB}$
(e) $\mathbf{AC}$ **(f)** $\mathbf{CA}$

2 Find the matrix **B** such that $\mathbf{AB} = \mathbf{I}$ where $\mathbf{A} = \begin{bmatrix} 8 & -10 \\ -3 & 4 \end{bmatrix}$. *Sections 4.6*

Test yourself ANSWERS

1 **(a)** $\begin{bmatrix} 13 & 2 \\ 1 & -6 \end{bmatrix}$; **(b)** $\begin{bmatrix} 4 & -1 \\ -4 & 7 \end{bmatrix}$; **(c)** $\begin{bmatrix} 21 & 0 \\ -7 & 8 \end{bmatrix}$;

(d) $\begin{bmatrix} 14 & -1 \\ -6 & 3 \end{bmatrix}$; **(e)** $\begin{bmatrix} 17 & -6 \\ -5 & 0 \end{bmatrix}$; **(f)** $\begin{bmatrix} 15 & 10 \\ 6 & 2 \end{bmatrix}$.

2 $\mathbf{B} = \begin{bmatrix} 2 & 5 \\ \frac{3}{2} & 4 \end{bmatrix}$.

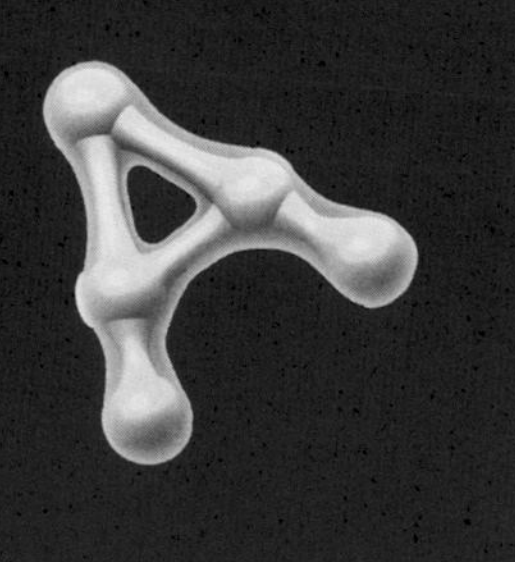

CHAPTER 5

Trigonometry

Learning objectives

After studying this chapter, you should be able to:

- recall and use the exact values of sine, cosine and tangent of any angle which is a multiple of 30° or a multiple of 45° and also the equivalent results when the angle is measured in radians
- find the general solutions, in degrees or radians, of trigonometric equations.

5.1 Exact values for sine, cosine and tangent of angles which are multiples of 30° and 45° and equivalent results for radian measure

Consider an equilateral triangle PQR of side 2. The angles of PQR are each 60° and PA is a line of symmetry which bisects both angle QPR and side QR.

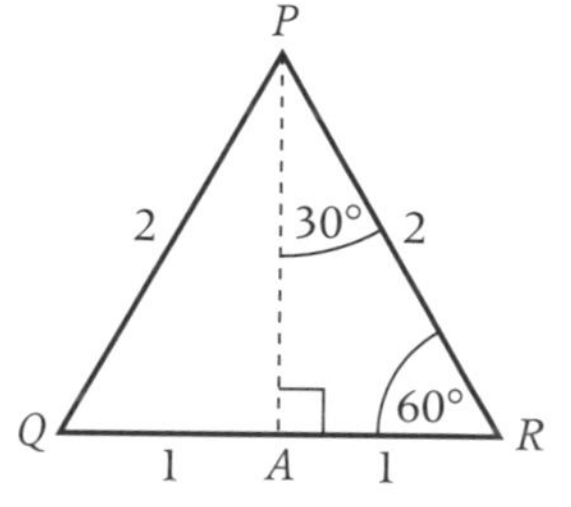

Using Pythagoras' theorem for triangle PAR leads to

$$PA^2 = 2^2 - 1^2 \Rightarrow PA = \sqrt{3}$$

Using basic trigonometry in triangle PAR we can now deduce that

$$\sin 30° = \frac{1}{2}, \qquad \cos 30° = \frac{\sqrt{3}}{2}, \qquad \tan 30° = \frac{1}{\sqrt{3}},$$

$$\sin 60° = \frac{\sqrt{3}}{2}, \qquad \cos 60° = \frac{1}{2}, \qquad \tan 60° = \sqrt{3}.$$

Note that $\sin 30° = \cos 60°$
and $\sin 60° = \cos 30$.
In general $\sin (90 - x)° = \cos x°$ $= \frac{b}{c}$.

c b $x°$ $(90 - x)°$

Consider a square of side 1, with the diagonal drawn. This forms an isosceles right-angled triangle of sides 1, 1 and $\sqrt{2}$.

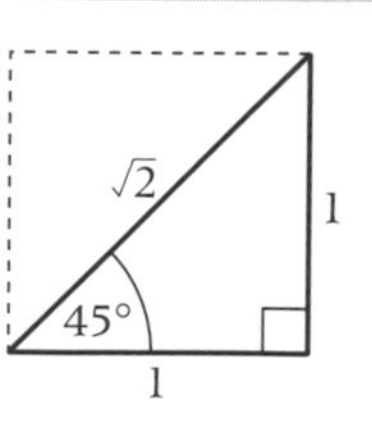

Using basic trigonometry in the triangle we can now deduce that

$$\sin 45° = \frac{1}{\sqrt{2}}, \qquad \cos 45° = \frac{1}{\sqrt{2}}, \qquad \tan 45° = 1.$$

You will recall from C2 that $360° = 2\pi$ radians which leads to $30° = \frac{\pi}{6}$ rads, $60° = \frac{\pi}{3}$ rads and $45° = \frac{\pi}{4}$ rads.

These results, along with some other exact results are summarised in the following table. They frequently appear in examination questions and should be memorised.

θ in degrees	θ in radians	$\sin\theta$	$\cos\theta$	$\tan\theta$
0	0	0	1	0
30	$\frac{\pi}{6}$	$\frac{1}{2}$	$\frac{\sqrt{3}}{2}$	$\frac{1}{\sqrt{3}}$
45	$\frac{\pi}{4}$	$\frac{1}{\sqrt{2}}$	$\frac{1}{\sqrt{2}}$	1
60	$\frac{\pi}{3}$	$\frac{\sqrt{3}}{2}$	$\frac{1}{2}$	$\sqrt{3}$
90	$\frac{\pi}{2}$	1	0	∞
180	π	0	-1	0
360	2π	0	1	0

You will need to use some of the following results from C2 chapter 4:

$\sin(360° + \theta) = \sin\theta$ $\sin(360° - \theta) = -\sin\theta$
$\sin(180° - \theta) = \sin\theta$ $\sin(180° + \theta) = -\sin\theta$
$\sin(-\theta) = -\sin\theta$

$\cos(360° + \theta) = \cos\theta$ $\cos(180° - \theta) = -\cos\theta$
$\cos(360° - \theta) = \cos\theta$ $\cos(180° + \theta) = -\cos\theta$
$\cos(-\theta) = \cos\theta$

$\tan(360° + \theta) = \tan\theta$ $\tan(360° - \theta) = -\tan\theta$
$\tan(180° + \theta) = \tan\theta$ $\tan(180° - \theta) = -\tan\theta$
$\tan(-\theta) = -\tan\theta$

Worked example 5.1

Find the exact value of $\cos\frac{5\pi}{6}$.

Solution

$$\cos\frac{5\pi}{6} = \cos\left(\pi - \frac{\pi}{6}\right) = -\cos\frac{\pi}{6}$$

$$\Rightarrow \quad \cos\frac{5\pi}{6} = -\frac{\sqrt{3}}{2}$$

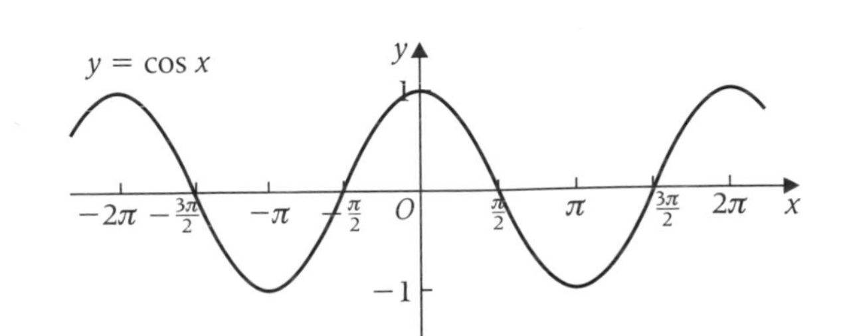

Worked example 5.2

Find the exact value of $\dfrac{\tan 135° - \sin 240°}{\cos 300°}$.

Solution

$\tan 135° = \tan(180 - 45)° = -\tan 45°$

$\sin 240° = \sin(180 + 60)° = -\sin 60°$

$\cos 300° = \cos(360 - 60)° = \cos 60°$

$$\frac{\tan 135° - \sin 240°}{\cos 300°} = \frac{-\tan 45° + \sin 60°}{\cos 60°}$$

$$= \frac{-1 + \dfrac{\sqrt{3}}{2}}{\dfrac{1}{2}} = \frac{\dfrac{1}{2}(\sqrt{3} - 2)}{\dfrac{1}{2}} = \sqrt{3} - 2.$$

$$\Rightarrow \quad \frac{\tan 135° - \sin 240°}{\cos 300°} = \sqrt{3} - 2$$

EXERCISE 5A

1 Find the exact values of the following:

(a) $\sin 300°$, **(b)** $\tan 120°$, **(c)** $\cos 225°$,

(d) $\sin 150°$, **(e)** $\tan 510°$, **(f)** $\cos 150°$,

(g) $\sin 270° + \tan 210° + \cos 420°$,

(h) $\sin(-315°) + \tan(-405°) - \cos(-120°)$.

2 Find the exact values of the following:

(a) $\sin \dfrac{2\pi}{3}$, **(b)** $\tan \dfrac{3\pi}{4}$, **(c)** $\cos \dfrac{7\pi}{6}$,

(d) $\sin\left(-\dfrac{\pi}{6}\right)$, **(e)** $\tan\left(-\dfrac{2\pi}{3}\right)$, **(f)** $\cos\left(-\dfrac{5\pi}{4}\right)$,

(g) $\sin \dfrac{11\pi}{6}$, **(h)** $\tan \dfrac{14\pi}{3}$, **(i)** $\cos \dfrac{8\pi}{3}$,

(j) $\sin(-2\pi)$, **(k)** $\tan\left(\dfrac{-7\pi}{6}\right)$, **(l)** $\cos(-7\pi)$.

3 Without using a calculator, verify that $\dfrac{\pi}{8}$ is a solution of the equation $\sin 3x = \cos x$.

4 Find the exact value of $\dfrac{3\tan\dfrac{5\pi}{4}}{\sin\dfrac{\pi}{3}} + \cos\left(-\dfrac{7\pi}{6}\right)$.

5 Given that $x = \dfrac{\pi}{4}$ is a solution of the equation

$3\sin^2 x - k\tan x = \dfrac{1}{3}$, find the value of the constant k.

6 Given that $\sin\left(-\dfrac{5\pi}{6}\right) + k\tan\dfrac{7\pi}{6} + \cos\dfrac{5\pi}{6} = 0$, find the value of the constant k.

5.2 General solutions of sin *x* = *k* and cos *x* = *k*, where −1 ⩽ *k* ⩽ 1

In C2 you solved equations of the form $\sin x = k$ and $\cos x = k$, for values of x within a specified interval, for example $-180° \leqslant x \leqslant 180°$ (when x is in degrees) and $-\pi \leqslant x \leqslant \pi$ (when x is in radians). In the remaining sections of this chapter you will be shown how to give your answers in more general terms which represent the infinite number of solutions which trigonometric equations have. This is called finding the general solution of a trigonometric equation.

To find the general solution of $\sin x = k$ or $\cos x = k$, where $-1 \leqslant k \leqslant 1$, you find two solutions lying in the interval $-180° \leqslant x \leqslant 180°$ and then, since sine and cosine are periodic with period 360°, you add $360n°$, where n is an integer, to each of the two solutions.

For x in radians you find the general solution of $\sin x = k$ and $\cos x = k$, where $-1 \leqslant k \leqslant 1$, by finding two solutions lying in the interval $-\pi \leqslant x \leqslant \pi$ and then adding $2n\pi$, where n is an integer, to each of the two solutions.

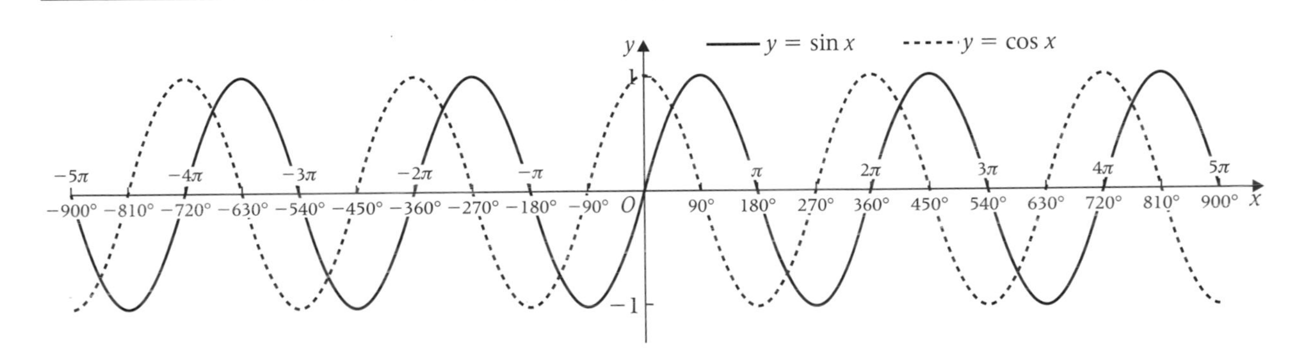

Worked example 5.3

Find the general solution, in degrees, of the equation $\cos\theta = \frac{1}{2}$.

Solution

$\cos\theta = \frac{1}{2}$

$\theta = \cos^{-1}(\frac{1}{2}) = 60°$

The other solution in the interval $-180° \leqslant \theta \leqslant 180°$ is $\theta = -60°$.

The general solution of $\cos\theta = \frac{1}{2}$ is

$\theta = 360n° + 60°$, $\theta = 360n° - 60°$.

These can be written in the combined form $\theta = 360n° \pm 60°$, where n is an integer.

Particular solutions can be found for a specified interval by choosing the appropriate integer values of n, for example if the interval was $-720° \leqslant \theta \leqslant 0°$ you would choose $n = -2, -1$ and 0 to get $-660°, -420°, -300°$ and $-60°$ as the four solutions in the interval.

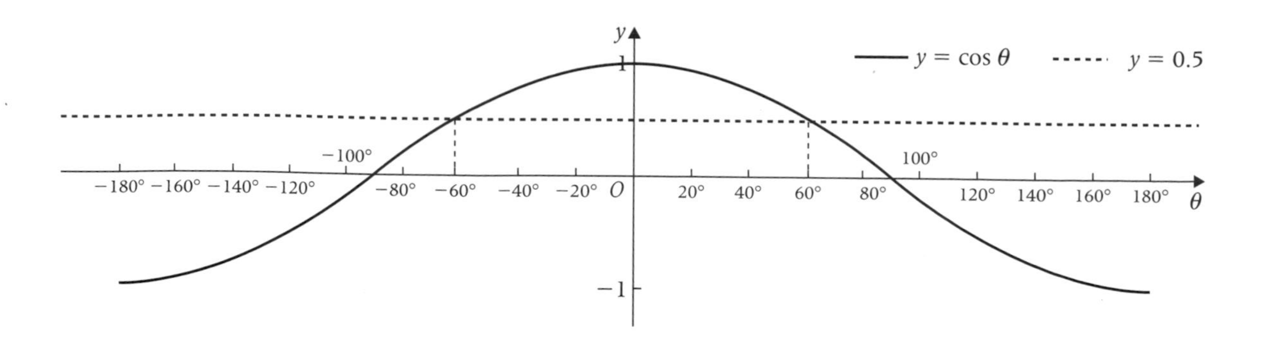

Worked example 5.4

Find the general solution, in radians, of the equation $\sin 2\theta = \frac{1}{\sqrt{2}}$.

Solution

Let $x = 2\theta$, so $\sin 2\theta = \frac{1}{\sqrt{2}}$ becomes

$\sin x = \frac{1}{\sqrt{2}}$.

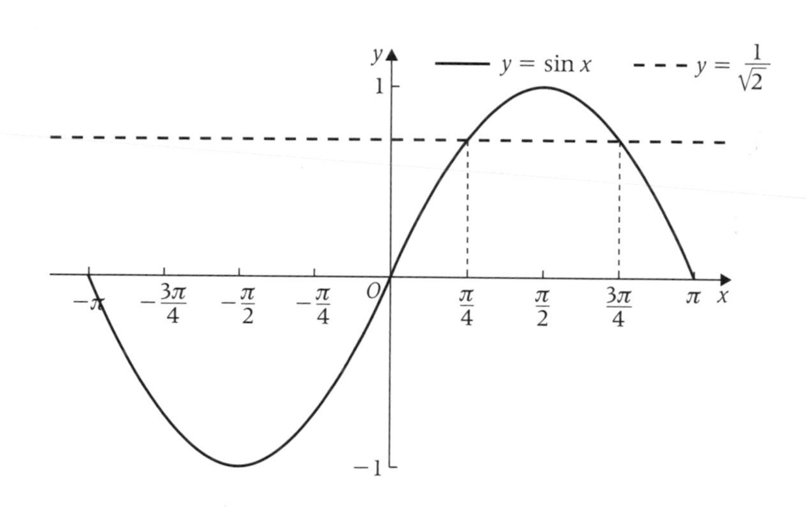

One solution is $x = \sin^{-1}\left(\frac{1}{\sqrt{2}}\right) = \frac{\pi}{4}$.

The other solution in the interval

$-\pi < x < \pi$ is $x = \pi - \frac{\pi}{4} = \frac{3\pi}{4}$.

The general solution of $\sin x = \frac{1}{\sqrt{2}}$ is

$$x = 2n\pi + \frac{\pi}{4}, \quad x = 2n\pi + \frac{3\pi}{4}.$$

$$\Rightarrow \quad 2\theta = 2n\pi + \frac{\pi}{4}, \quad 2\theta = 2n\pi + \frac{3\pi}{4}$$

The full general solution should be found for 2θ before dividing by 2 to find the general values of θ.

which leads to the general solution $\theta = n\pi + \frac{\pi}{8}$, $\theta = n\pi + \frac{3\pi}{8}$, where n is an integer.

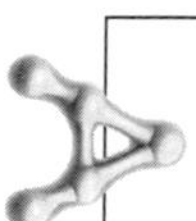

The general solution of $\cos x = \cos \alpha$ is $x = 2n\pi \pm \alpha$ (or $x = (360n \pm \alpha)°$, if x is in degrees), where n is an integer.

The general solution of $\sin x = \sin \alpha$ is $x = 2n\pi + \alpha$, $2n\pi + \pi - \alpha$ (or $x = (360n + \alpha)°$, $(360n + 180 - \alpha)°$, if x is in degrees), where n is an integer.

EXERCISE 5B

1 Find the general solutions, in degrees, of these equations:

(a) $\sin x = \frac{1}{\sqrt{2}}$, (b) $\cos x = 0$,

(c) $\sin 2x = -\frac{\sqrt{3}}{2}$, (d) $\cos 4x = \frac{1}{2}$,

(e) $\sin 2x = 0.6$, (f) $1 + \cos 3x = 0.3$,

(g) $\sin(x + 40°) = \sin 50°$ (h) $\cos(50° - 2x) = \cos 20°$.

2 Find the general solutions, in radians, of these equations:

(a) $\cos x = \frac{1}{\sqrt{2}}$, (b) $\sin x = 0$,

(c) $\cos 4x = \frac{\sqrt{3}}{2}$, (d) $\sin 2x = -\frac{1}{2}$,

(e) $\cos 2x = 0.6$, (f) $1 + \sin(3x + 1) = 0.3$,

(g) $\cos\left(x - \frac{\pi}{2}\right) = \cos\frac{\pi}{3}$, (h) $\sin\left(\frac{\pi}{4} - 2x\right) = 1$.

3 Find the general solution, in radians, of the equation $2\sin x \cos x - \cos x = 0$.

4 Find the general solution, in degrees, of the equation $2\cos^2 x = \cos x$.

5 Find the general solution, in radians, of the equation $2\cos^2 x = 1$.

6 Find the general solution, in degrees, of the equation $4\sin^2 2x = 3$.

5.3 General solution of tan x = k

In C2 you solved equations of the form $\tan x = k$ for values of x within a specified interval. You saw that the graph of $y = \tan x$ was periodic with period 180°. You can use this periodicity to find the general solution of $\tan x = k$.

To find the general solution of $\tan x = k$, where k is a constant, you find the solution lying in the interval $-90° \leqslant x \leqslant 90°$ and then, since tan is periodic with period 180°, you add $180n°$, where n is an integer, to the solution.

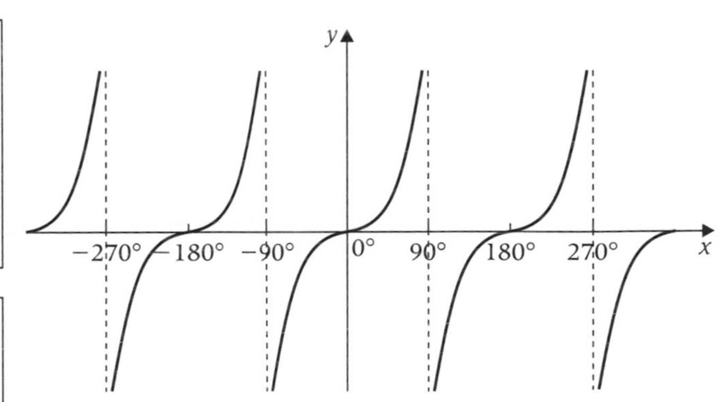

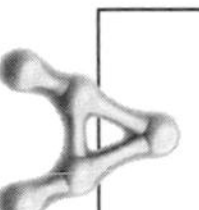

For x in radians you find the general solution of $\tan x = k$ by finding the solution lying in the interval $-\frac{\pi}{2} \leqslant x \leqslant \frac{\pi}{2}$ then adding $n\pi$, where n is an integer, to the solution.

Worked example 5.5

Find the general solution, in degrees, of the equation $4 \tan 2\theta + 3 = 0$.

Solution

$4 \tan 2\theta + 3 = 0 \quad \Rightarrow \quad \tan 2\theta = -0.75.$

$2\theta = \tan^{-1}(-0.75) = -36.869...°.$

The general solution for 2θ is $2\theta = 180n° - 36.869...°$

$\Rightarrow \quad \theta = 90n° - 18.4°$ (to 3 sf), where n is an integer.

Worked example 5.6

Find the general solution, in radians, of the equation $\tan^2(1 - 2\theta) = 3$.

Solution

Let $x = 1 - 2\theta$ so $\tan^2(1 - 2\theta) = 3$ becomes $\tan^2 x = 3$.

$\tan^2 x = 3 \quad \Rightarrow \quad \tan x = \pm\sqrt{3}$

When $\tan x = \sqrt{3}$, $x = \tan^{-1}\sqrt{3} = \dfrac{\pi}{3}$.

The general solution of $\tan x = \sqrt{3}$ is $x = n\pi + \dfrac{\pi}{3}$.

When $\tan x = -\sqrt{3}$, $x = \tan^{-1}(-\sqrt{3}) = -\dfrac{\pi}{3}$.

The general solution of $\tan x = -\sqrt{3}$ is $x = n\pi - \dfrac{\pi}{3}$.

The general solution of $\tan^2 x = 3$ is $x = n\pi + \dfrac{\pi}{3}$ and $x = n\pi - \dfrac{\pi}{3}$.

$\Rightarrow \quad 1 - 2\theta = n\pi + \dfrac{\pi}{3}$ and $1 - 2\theta = n\pi - \dfrac{\pi}{3}$.

$\Rightarrow \quad \theta = \dfrac{1}{2} - \dfrac{n\pi}{2} - \dfrac{\pi}{6}$ and $\theta = \dfrac{1}{2} - \dfrac{n\pi}{2} + \dfrac{\pi}{6}$

So the general solution of $\tan^2(1 - 2\theta) = 3$ is

$\theta = \dfrac{1}{2} - \dfrac{n\pi}{2} \pm \dfrac{\pi}{6}$, where n is an integer.

The general solution of $\tan x = \tan \alpha$ is $x = n\pi + \alpha$ (or $x = 180n + \alpha)°$, if x is in degrees), where n is an integer.

EXERCISE 5C

1 Find the general solutions, in degrees, of these equations:

(a) $\tan x = \tan 20°$, **(b)** $\tan x = -1$,

(c) $\tan 2x = \tan 70°$, **(d)** $\tan 4x = \tan(-40°)$,

(e) $\tan (x - 75°) = -\tan 50°$, **(f)** $2 \tan (50° - 2x) = 5$.

2 Find the general solutions, in radians, of these equations:

(a) $\tan x = 1$, **(b)** $\tan x = -\sqrt{3}$,

(c) $\tan 4x + 1 = 0$, **(d)** $\tan 2x = \frac{1}{\sqrt{3}}$,

(e) $\tan (2x - \frac{\pi}{2}) = \tan \frac{\pi}{3}$, **(f)** $\tan \left(\frac{\pi}{4} - 2x\right) = -\tan \frac{\pi}{3}$.

3 Find the general solution, in radians, of the equation $3 \tan^2 x = 1$.

4 Find the general solution, in degrees, of the equation $(\tan x + 1)(\tan x - 3) = 0$.

5 Find the general solution, in radians, of the equation $\tan 2x = \tan x$.

Hint: Use Key point 8 to find $2x$ in terms of x.

6 Find the general solution, in degrees, of the equation $\tan 4x = \tan (x + 42°)$.

MIXED EXERCISE

In this exercise you may need to use the two identities $\cos^2 \theta + \sin^2 \theta \equiv 1$ and $\tan \theta \equiv \frac{\sin \theta}{\cos \theta}$ which you first considered in C2 chapter 6.

1 (a) Given that $3 \sin x = \sqrt{3} \cos x$, write down the value of $\tan x$.

(b) Hence, or otherwise, find the general solution, in degrees, of $3 \sin x = \sqrt{3} \cos x$.

2 (a) Show that the equation $\cos x + 1 = 2 \sin^2 x$ can be written in the form $2 \cos^2 x + \cos x - 1 = 0$.

(b) Hence find the general solution, in radians, of the equation $\cos x + 1 = 2 \sin^2 x$.

3 (a) Write down the exact values of $\sin \frac{\pi}{6}$, $\cos \frac{\pi}{6}$ and $\tan \frac{\pi}{6}$.

(b) It is given that x satisfies the equation $3 \sin^2 x = \cos^2 x$. By first using a trigonometrical identity to simplify this equation, find the general solution in radians, of the equation $3 \sin^2 x = \cos^2 x$. [A adapted]

4 It is given that x satisfies the equation $2\cos^2 x = 2 + \sin x$.

(a) Use an appropriate trigonometrical identity to show that $2\sin^2 x + \sin x = 0$.

(b) Solve this quadratic equation and hence find the general solution, in radians, of the equation $2\cos^2 x = 2 + \sin x$.

(c) Hence find the general solution, in radians, of the equation $2\cos^2 3\theta = 2 + \sin 3\theta$. [A adapted]

5 Find the general solution, in degrees, of the equation $\sin\theta\tan\theta = \cos\theta$.

6 Find the general solution, in degrees, of the equation $(\sin x - \cos x)(4\sin^2 x - 3) = 0$.

7 Find the general solution, in radians, of the equation $2\sin^2\theta + 5\cos\theta + 1 = 0$. [A]

Key point summary

1

θ in degrees	θ in radians	$\sin\theta$	$\cos\theta$	$\tan\theta$
0	0	0	1	0
30	$\frac{\pi}{6}$	$\frac{1}{2}$	$\frac{\sqrt{3}}{2}$	$\frac{1}{\sqrt{3}}$
45	$\frac{\pi}{4}$	$\frac{1}{\sqrt{2}}$	$\frac{1}{\sqrt{2}}$	1
60	$\frac{\pi}{3}$	$\frac{\sqrt{3}}{2}$	$\frac{1}{2}$	$\sqrt{3}$
90	$\frac{\pi}{2}$	1	0	∞
180	π	0	-1	0
360	2π	0	1	0

p53

2 To find the general solution of $\sin x = k$ or $\cos x = k$, where $-1 \leqslant k \leqslant 1$, you find two solutions lying in the interval $-180° \leqslant x \leqslant 180°$ and then, since sine and cosine are periodic with period $360°$, you add $360n°$, where n is an integer, to each of the two solutions. *p55*

3 For x in radians you find the general solution of $\sin x = k$ and $\cos x = k$, where $-1 \leqslant k \leqslant 1$ by finding two solutions lying in the interval $-\pi \leqslant x \leqslant \pi$ and then adding $2n\pi$, where n is an integer, to each of the two solutions. *p55*

4 The general solution of $\cos x = \cos\alpha$ is $x = 2n\pi \pm \alpha$ (or $x = (360n \pm \alpha)°$, if x is in degrees), where n is an integer. *p56*

5 The general solution of $\sin x = \sin\alpha$ is $x = 2n\pi + \alpha$, $2n\pi + \pi - \alpha$ (or $x = (360n + \alpha)°$, $(360n + 180 - \alpha)°$, if x is in degrees), where n is an integer. *p57*

6 To find the general solution of $\tan x = k$, where k is a constant, you find the solution lying in the interval $-90° \leqslant x \leqslant 90°$ and then, since tan is periodic with period 180°, you add $180n°$, where n is an integer, to the solution. p57

7 For x in radians you find the general solution of $\tan x = k$ by finding the solution lying in the interval $-\frac{\pi}{2} \leqslant x \leqslant \frac{\pi}{2}$ then adding $n\pi$, where n is an integer, to the solution. p57

8 The general solution of $\tan x = \tan \alpha$ is $x = n\pi + \alpha$ (or $x = (180n + \alpha)°$, if x is in degrees), where n is an integer. p58

5

Test yourself

Test yourself	What to review
1 Find the exact value of $\dfrac{\tan \frac{11\pi}{6}}{\sin \frac{\pi}{3} \cos \frac{9\pi}{4}} + 2 \sin\left(-\frac{5\pi}{4}\right)$.	*Section 5.1*
2 Find the general solution of the equation $8 \sin(34° - 2x) = 3$, giving your answer in degrees to the nearest degree.	*Section 5.2*
3 Find the general solution, in radians, of the equation $\cos\left(2x + \frac{\pi}{4}\right) = -\frac{\sqrt{3}}{2}$.	*Section 5.2*
4 Find the general solution, in radians, of the equation $4 + \tan 4x = 3$.	*Section 5.3*

Test yourself ANSWERS

Note: Correct answers for general solutions may be in a different form to those given below; for example $(6 - 180n)°$ is a valid correct alternative to $(180n + 6)°$ in question 2.

1 $\frac{\sqrt{2}}{3}$.

2 $x = (180n + 6)°$, $x = (180n - 62)°$, where n is an integer.

3 $x = n\pi + \frac{7\pi}{24}$, $x = n\pi - \frac{13\pi}{24}$.

4 $x = \frac{n\pi}{4} - \frac{\pi}{16}$.

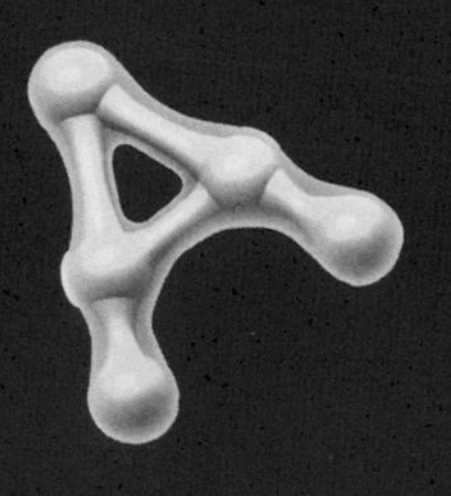

CHAPTER 6

Matrix transformations

Learning objectives

After studying this chapter, you should be able to:

- write down a transformation in matrix and algebraic form
- recognise the matrices associated with common transformations
- form matrices associated with composite transformations.

6.1 Introduction

You will have studied various types of transformations in the past. You may not have been aware that transformations can be expressed in both algebraic and matrix form as you will see. In this chapter you will be considering various common transformations and their associated matrices.

(i) stretches
(ii) enlargements
(iii) rotations
(iv) reflections

Worked example 6.1

Consider a reflection in the line $y = x$. Write down the algebraic and matrix forms of this transformation.

Solution

Firstly, consider a point $P(x, y)$ and its reflection $P'(x', y')$ in the line $y = x$.

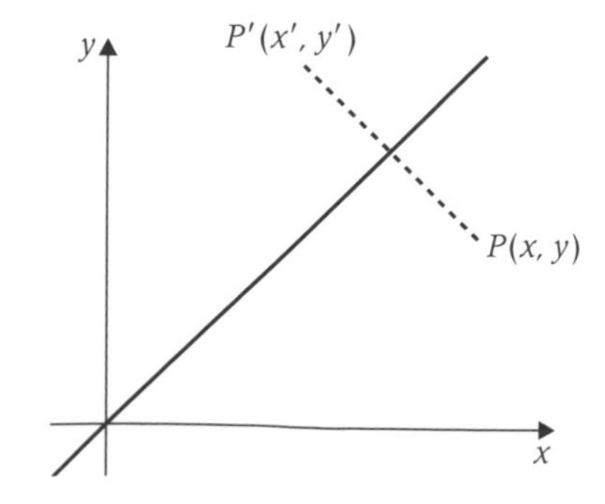

P is the **object** and the reflection *P′* is called the **image**.
You can think of (x, y) as the *old* coordinates and (x', y') as the *new* coordinates.

It should be clear to you from your previous knowledge of reflections in the line $y = x$ that the image (x, y) is reflected to (y, x).

This means that the *new* x-coordinate is the same as the *old* y-coordinate and the *new* y-coordinate is the same as the *old* x-coordinate.

This can be expressed algebraically as:

$$x' = y$$
$$y' = x$$

This is the algebraic form mentioned at the beginning of the section.

This can also be expressed as a matrix equation as:

$$\begin{bmatrix} x' \\ y' \end{bmatrix} = \begin{bmatrix} 0 & 1 \\ 1 & 0 \end{bmatrix} \begin{bmatrix} x \\ y \end{bmatrix}$$

This is the matrix form mentioned at the beginning of the section.

The matrix $\begin{bmatrix} 0 & 1 \\ 1 & 0 \end{bmatrix}$ is the **transformation matrix** associated with a reflection in the line $y = x$.

6

6.2 Transformation matrices

All **linear transformations** of the x–y plane, where the origin remains fixed, can be described both in matrix form and algebraic form.

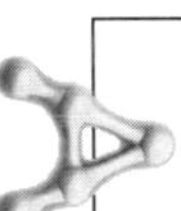

A linear transformation that changes/transforms point $P(x, y)$ into point $P'(x', y')$ can be written as:

$$\begin{bmatrix} x' \\ y' \end{bmatrix} = \begin{bmatrix} a & b \\ c & d \end{bmatrix} \begin{bmatrix} x \\ y \end{bmatrix} \quad \textbf{matrix form}$$

or $x' = ax + by$ **algebraic form**

$y' = cx + dy$

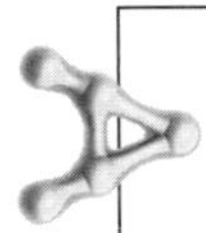

The matrix $\begin{bmatrix} a & b \\ c & d \end{bmatrix}$ is called a transformation matrix.

Worked example 6.2

A transformation is given by $\begin{bmatrix} x' \\ y' \end{bmatrix} = \begin{bmatrix} 3 & 1 \\ 4 & 5 \end{bmatrix} \begin{bmatrix} x \\ y \end{bmatrix}$.

(a) Find the images of the points $P(2, 5)$ and $Q(-1, 3)$ under this transformation.

(b) Find the point R whose image is $(7, 24)$.

Solution

(a) $\begin{bmatrix} 3 & 1 \\ 4 & 5 \end{bmatrix}\begin{bmatrix} 2 \\ 5 \end{bmatrix} = \begin{bmatrix} 11 \\ 33 \end{bmatrix} \Rightarrow$ image of P is $(11, 33)$

$\begin{bmatrix} 3 & 1 \\ 4 & 5 \end{bmatrix}\begin{bmatrix} -1 \\ 3 \end{bmatrix} = \begin{bmatrix} 0 \\ 11 \end{bmatrix} \Rightarrow$ image of Q is $(0, 11)$

(b) Let the point R have coordinates (a, b)

Now $\begin{bmatrix} 3 & 1 \\ 4 & 5 \end{bmatrix}\begin{bmatrix} a \\ b \end{bmatrix} = \begin{bmatrix} 7 \\ 24 \end{bmatrix}$.

Multiplying out gives:

$$3a + b = 7$$

and $4a + 5b = 24$.

Solving the simultaneous equations gives $a = 1$ and $b = 4$.

So R has coordinates $(1, 4)$.

Worked example 6.3

A transformation transforms the point $(3, 2)$ to $(11, 2)$ and the point $(2, 5)$ to $(22, 5)$. Find the transformation matrix associated with this transformation.

Solution

Let the transformation matrix be $\begin{bmatrix} a & b \\ c & d \end{bmatrix}$.

Then $\begin{bmatrix} a & b \\ c & d \end{bmatrix}\begin{bmatrix} 3 \\ 2 \end{bmatrix} = \begin{bmatrix} 11 \\ 2 \end{bmatrix} \Rightarrow 3a + 2b = 11$ [1]

and $3c + 2d = 2$ [2]

Also $\begin{bmatrix} a & b \\ c & d \end{bmatrix}\begin{bmatrix} 2 \\ 5 \end{bmatrix} = \begin{bmatrix} 22 \\ 5 \end{bmatrix} \Rightarrow 2a + 5b = 22$ [3]

and $2c + 5d = 5$ [4]

To find the values of a and b you can solve equations [1] and [3] simultaneously:

$3a + 2b = 11$ [1]

$2a + 5b = 22$ [3]

$\Rightarrow$ $15a + 10b = 55$

$4a + 10b = 44$

$\Rightarrow$ $11a = 11$

$\Rightarrow$ $a = 1$ and $b = 4$

To find the values of c and d you can solve equations [2] and [4] simultaneously:

$$3c + 2d = 2 \qquad [2]$$
$$2c + 5d = 5 \qquad [4]$$
$$\Rightarrow \quad 15c + 10d = 10$$
$$4c + 10d = 10$$
$$\Rightarrow \quad 11c = 0$$
$$\Rightarrow \quad c = 0 \quad \text{and} \quad d = 1$$

Thus, the transformation matrix that is associated with the given transformation is $\begin{bmatrix} 1 & 4 \\ 0 & 1 \end{bmatrix}$.

6.3 Matrices associated with common transformations

A good method of observing the geometric effect of a transformation matrix is to calculate its effect on the unit square with vertices at $O(0, 0)$, A(0, 1), B(1, 1) and C(1, 0).

6.4 Stretches and enlargements

Worked example 6.4

Describe the geometric effect of the transformation associated with the following matrices:

(a) $\begin{bmatrix} 3 & 0 \\ 0 & 1 \end{bmatrix}$, **(b)** $\begin{bmatrix} 1 & 0 \\ 0 & 5 \end{bmatrix}$.

Solution

Transformation matrix.

Old coordinates.

(a) $O(0, 0)$ $\begin{bmatrix} x' \\ y' \end{bmatrix} = \begin{bmatrix} 3 & 0 \\ 0 & 1 \end{bmatrix}\begin{bmatrix} 0 \\ 0 \end{bmatrix} = \begin{bmatrix} 0 \\ 0 \end{bmatrix}$ ← New coordinates.

so $O(0, 0) \rightarrow O'(0, 0)$.

$A(0, 1)$ $\begin{bmatrix} x' \\ y' \end{bmatrix} = \begin{bmatrix} 3 & 0 \\ 0 & 1 \end{bmatrix}\begin{bmatrix} 0 \\ 1 \end{bmatrix} = \begin{bmatrix} 0 \\ 1 \end{bmatrix}$

so $A(0, 1) \rightarrow A'(0, 1)$.

$B(1, 1)$ $\begin{bmatrix} x' \\ y' \end{bmatrix} = \begin{bmatrix} 3 & 0 \\ 0 & 1 \end{bmatrix}\begin{bmatrix} 1 \\ 1 \end{bmatrix} = \begin{bmatrix} 3 \\ 1 \end{bmatrix}$

so $B(1, 1) \rightarrow B'(3, 1)$.

$C(1, 0)$ $\begin{bmatrix} x' \\ y' \end{bmatrix} = \begin{bmatrix} 3 & 0 \\ 0 & 1 \end{bmatrix}\begin{bmatrix} 1 \\ 0 \end{bmatrix} = \begin{bmatrix} 3 \\ 0 \end{bmatrix}$

so $C(1, 0) \rightarrow C'(3, 0)$.

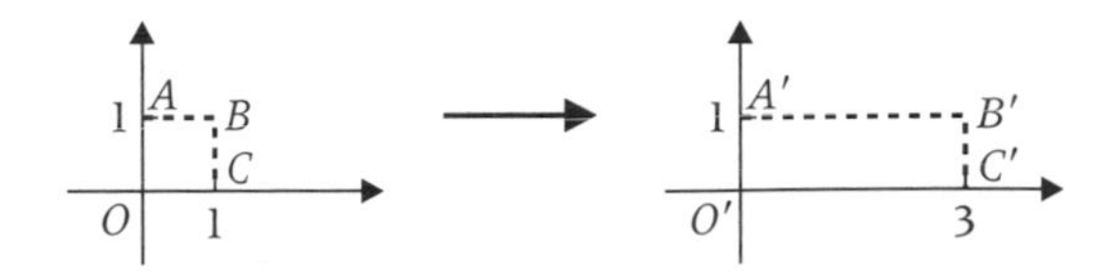

So $\begin{bmatrix} 3 & 0 \\ 0 & 1 \end{bmatrix}$ represents a one-way stretch in the x-direction with scale factor 3.

A stretch of this type is referred to as a one-way stretch as it takes place in one direction only.

(b) $O(0, 0)$ $\begin{bmatrix} x' \\ y' \end{bmatrix} = \begin{bmatrix} 1 & 0 \\ 0 & 5 \end{bmatrix}\begin{bmatrix} 0 \\ 0 \end{bmatrix} = \begin{bmatrix} 0 \\ 0 \end{bmatrix}$

so $O(0, 0) \rightarrow O'(0, 0)$.

$A(0, 1)$ $\begin{bmatrix} x' \\ y' \end{bmatrix} = \begin{bmatrix} 1 & 0 \\ 0 & 5 \end{bmatrix}\begin{bmatrix} 0 \\ 1 \end{bmatrix} = \begin{bmatrix} 0 \\ 5 \end{bmatrix}$

so $A(0, 1) \rightarrow A'(0, 5)$.

$B(1, 1)$ $\begin{bmatrix} x' \\ y' \end{bmatrix} = \begin{bmatrix} 1 & 0 \\ 0 & 5 \end{bmatrix}\begin{bmatrix} 1 \\ 1 \end{bmatrix} = \begin{bmatrix} 1 \\ 5 \end{bmatrix}$

so $B(1, 1) \rightarrow B'(1, 5)$.

$C(1, 0)$ $\begin{bmatrix} x' \\ y' \end{bmatrix} = \begin{bmatrix} 1 & 0 \\ 0 & 5 \end{bmatrix}\begin{bmatrix} 1 \\ 0 \end{bmatrix} = \begin{bmatrix} 1 \\ 0 \end{bmatrix}$

so $C(1, 0) \rightarrow C'(1, 0)$.

So $\begin{bmatrix} 1 & 0 \\ 0 & 5 \end{bmatrix}$ represents a one-way stretch in the y-direction of scale factor 5.

The results from the previous example can be generalised to:

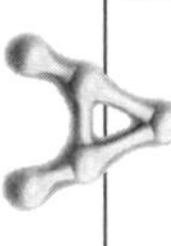

A one-way stretch in the x-direction of scale factor k is given by:

$$\begin{bmatrix} x' \\ y' \end{bmatrix} = \begin{bmatrix} k & 0 \\ 0 & 1 \end{bmatrix}\begin{bmatrix} x \\ y \end{bmatrix}.$$

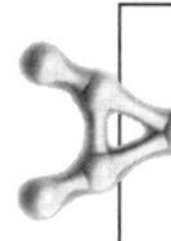

A one-way stretch in the y-direction of scale factor k is given by:

$$\begin{bmatrix} x' \\ y' \end{bmatrix} = \begin{bmatrix} 1 & 0 \\ 0 & k \end{bmatrix}\begin{bmatrix} x \\ y \end{bmatrix}.$$

Stretches can also take place in **both directions**. These are known as **two-way stretches**.

For example, it is not difficult to see that the matrix $\begin{bmatrix} 4 & 0 \\ 0 & 3 \end{bmatrix}$ represents a stretch in the x-direction of scale factor 4 and a stretch in the y-direction of scale factor 3.

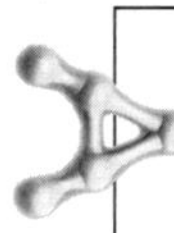

A two-way stretch of scale factor a in the x-direction and scale factor b in the y-direction is given by:

$$\begin{bmatrix} x' \\ y' \end{bmatrix} = \begin{bmatrix} a & 0 \\ 0 & b \end{bmatrix}\begin{bmatrix} x \\ y \end{bmatrix}.$$

Worked example 6.5

Describe the geometrical effect of the transformation given algebraically by

$$x' = 2x, \quad y' = 2y$$

by applying a suitable matrix to the unit square and illustrating the effect graphically.

Solution

Firstly, the required matrix is $\begin{bmatrix} 2 & 0 \\ 0 & 2 \end{bmatrix}$.

$$\begin{bmatrix} x' \\ y' \end{bmatrix} = \begin{bmatrix} 2 & 0 \\ 0 & 2 \end{bmatrix}\begin{bmatrix} x \\ y \end{bmatrix}$$

(a) $O(0, 0)$ $\quad \begin{bmatrix} x' \\ y' \end{bmatrix} = \begin{bmatrix} 2 & 0 \\ 0 & 2 \end{bmatrix}\begin{bmatrix} 0 \\ 0 \end{bmatrix} = \begin{bmatrix} 0 \\ 0 \end{bmatrix}$

so $O(0, 0) \to O'(0, 0)$.

$A(0, 1)$ $\quad \begin{bmatrix} x' \\ y' \end{bmatrix} = \begin{bmatrix} 2 & 0 \\ 0 & 2 \end{bmatrix}\begin{bmatrix} 0 \\ 1 \end{bmatrix} = \begin{bmatrix} 0 \\ 2 \end{bmatrix}$

so $A(0, 1) \to A'(0, 2)$.

$B(1, 1)$ $\quad \begin{bmatrix} x' \\ y' \end{bmatrix} = \begin{bmatrix} 2 & 0 \\ 0 & 2 \end{bmatrix}\begin{bmatrix} 1 \\ 1 \end{bmatrix} = \begin{bmatrix} 2 \\ 2 \end{bmatrix}$

so $B(1, 1) \to B'(2, 2)$.

$C(1, 0)$ $\quad \begin{bmatrix} x' \\ y' \end{bmatrix} = \begin{bmatrix} 2 & 0 \\ 0 & 2 \end{bmatrix}\begin{bmatrix} 1 \\ 0 \end{bmatrix} = \begin{bmatrix} 2 \\ 0 \end{bmatrix}$

so $C(1, 0) \to C'(2, 0)$.

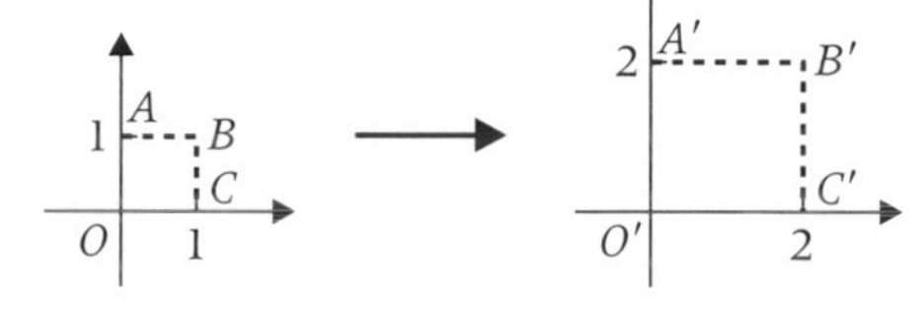

So $\begin{bmatrix} 2 & 0 \\ 0 & 2 \end{bmatrix}$ represents an enlargement of scale factor 2 with the centre at the origin.

In general,

> An enlargement of scale factor k with centre the origin is given by:
>
> $$\begin{bmatrix} x' \\ y' \end{bmatrix} = \begin{bmatrix} k & 0 \\ 0 & k \end{bmatrix}\begin{bmatrix} x \\ y \end{bmatrix}.$$

An enlargement is a special case of a two-way stretch where the scale factors in each direction are the same.

EXERCISE 6A

1 A transformation is given by $\begin{bmatrix} x' \\ y' \end{bmatrix} = \begin{bmatrix} 2 & 1 \\ 3 & 2 \end{bmatrix}\begin{bmatrix} x \\ y \end{bmatrix}$.

(a) Find the images of the points $A(0, 2)$, $B(3, 5)$ and $C(-4, 2)$ under this transformation.

(b) Under the transformation, the point D has image $(0, 2)$. Find the coordinates of point D.

2 A transformation is given by $x' = x$ and $y' = y - 2x$.

(a) Find the images of the points $P(3, 0)$, $Q(-4, 1)$ and $R(0, 6)$ under this transformation.

(b) Under this transformation, the point S has image $(4, -3)$. Find the coordinates of point S.

3 A transformation transforms the point $(3, 5)$ to $(7, 4)$ and the point $(-2, 4)$ to $(10, -10)$. Find the transformation matrix associated with this transformation.

4 Write down the matrices associated with the following transformations:

(a) an enlargement with scale factor 4, centre the origin,

(b) a stretch in the x-direction of scale factor 3,

(c) a stretch in the y-direction, scale factor 6,

(d) an enlargement with scale factor 8, centre the origin,

(e) an enlargement with scale factor -2, centre the origin,

(f) a stretch in the x-direction scale factor 4 and a stretch in the y-direction scale factor 2.

5 Describe the transformation associated with the following matrices:

(a) $\begin{bmatrix} 3 & 0 \\ 0 & 1 \end{bmatrix}$, **(b)** $\begin{bmatrix} 1 & 0 \\ 0 & 7 \end{bmatrix}$,

(c) $\begin{bmatrix} 6 & 0 \\ 0 & 7 \end{bmatrix}$, **(d)** $\begin{bmatrix} 5 & 0 \\ 0 & 5 \end{bmatrix}$,

(e) $\begin{bmatrix} 1 & 0 \\ 0 & 9 \end{bmatrix}$, **(f)** $\begin{bmatrix} -1 & 0 \\ 0 & -1 \end{bmatrix}$,

(g) $\begin{bmatrix} -3 & 0 \\ 0 & 1 \end{bmatrix}$, **(h)** $\begin{bmatrix} -4 & 0 \\ 0 & 9 \end{bmatrix}$,

(i) $\begin{bmatrix} 10 & 0 \\ 0 & 10 \end{bmatrix}$, **(j)** $\begin{bmatrix} 1 & 0 \\ 0 & -4 \end{bmatrix}$.

6 Find the image of the point $(3, 4)$ under the following transformations:

(a) an enlargement with scale factor 3, centre the origin,

(b) a stretch in the x-direction, scale factor 2,

(c) a stretch in the y-direction, scale factor 5,

(d) a stretch in the x-direction with scale factor -4.

6.5 Rotations about the origin

Suppose you want to find the matrix associated with a **rotation** through an **angle θ anticlockwise about the origin**.

In order to find this matrix you can consider the effect of the rotation on the two points $(1, 0)$ and $(0, 1)$.

In fact you can use this method to find the matrix associated with any transformation.

Find the image of (1, 0):

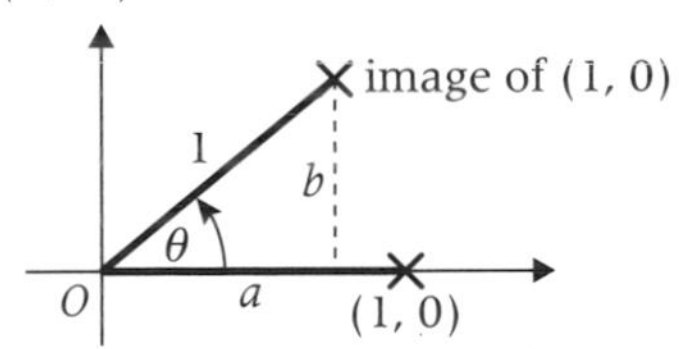

It should be clear from the diagram that

$$\cos\theta = \frac{a}{1} \quad \Rightarrow \quad a = \cos\theta$$

and $\sin\theta = \frac{b}{1} \quad \Rightarrow \quad b = \sin\theta.$

So the image of (1, 0) under the rotation is $(\cos\theta, \sin\theta)$. Written in column vector form (matrix form), this is

$$\begin{bmatrix} 1 \\ 0 \end{bmatrix} \rightarrow \begin{bmatrix} \cos\theta \\ \sin\theta \end{bmatrix}.$$

Find the image of (0, 1):

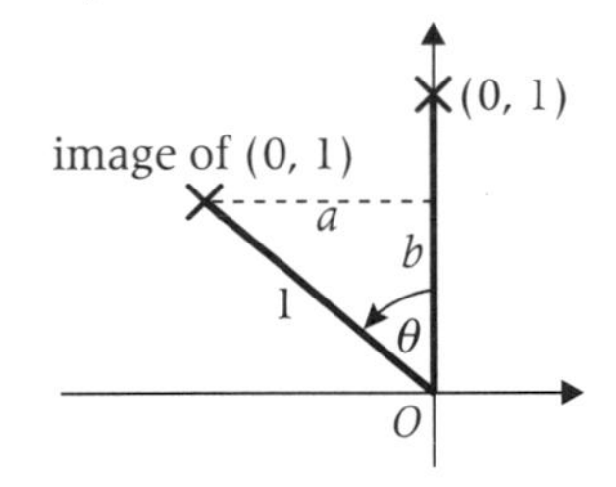

It should be clear from the diagram that

$$\sin\theta = \frac{a}{1} \quad \Rightarrow \quad a = \sin\theta$$

and $\cos\theta = \frac{b}{1} \quad \Rightarrow \quad b = \cos\theta.$

The x-coordinate of the image is negative.

So the image of (1, 0) under the rotation is $(-\sin\theta, \cos\theta)$. Written in column vector form (matrix form), this is

$$\begin{bmatrix} 0 \\ 1 \end{bmatrix} \rightarrow \begin{bmatrix} -\sin\theta \\ \cos\theta \end{bmatrix}.$$

Putting these two results together gives the rotation matrix as

$$\begin{bmatrix} \cos\theta & -\sin\theta \\ \sin\theta & \cos\theta \end{bmatrix}.$$

↑ Image of (1, 0). ↑ Image of (0, 1).

The images must be written down in this order.

So,

> A rotation through angle θ anticlockwise about the origin is given by:
>
> $$\begin{bmatrix} x' \\ y' \end{bmatrix} = \begin{bmatrix} \cos\theta & -\sin\theta \\ \sin\theta & \cos\theta \end{bmatrix} \begin{bmatrix} x \\ y \end{bmatrix}.$$

This matrix is in the formulae booklet provided in the examination.

Worked example 6.6

Find the matrix that generates the following rotations:

(a) A rotation of 30° anticlockwise about the origin,

(b) A rotation of 120° anticlockwise about the origin,

(c) A rotation of 45° clockwise about the origin.

Solution

(a) Substitute $\theta = 30°$ in the standard rotation matrix:

$$\begin{bmatrix} \cos\theta & -\sin\theta \\ \sin\theta & \cos\theta \end{bmatrix} = \begin{bmatrix} \cos 30° & -\sin 30° \\ \sin 30° & \cos 30° \end{bmatrix}$$

$$= \begin{bmatrix} \frac{\sqrt{3}}{2} & -\frac{1}{2} \\ \frac{1}{2} & \frac{\sqrt{3}}{2} \end{bmatrix} = \frac{1}{2}\begin{bmatrix} \sqrt{3} & -1 \\ 1 & \sqrt{3} \end{bmatrix}$$

(b) Substitute $\theta = 120°$ in the standard rotation matrix:

$$\begin{bmatrix} \cos\theta & -\sin\theta \\ \sin\theta & \cos\theta \end{bmatrix} = \begin{bmatrix} \cos 120° & -\sin 120° \\ \sin 120° & \cos 120° \end{bmatrix}$$

$$= \begin{bmatrix} -\frac{1}{2} & -\frac{\sqrt{3}}{2} \\ \frac{\sqrt{3}}{2} & -\frac{1}{2} \end{bmatrix} = -\frac{1}{2}\begin{bmatrix} 1 & \sqrt{3} \\ -\sqrt{3} & 1 \end{bmatrix}$$

(c) Substitute $\theta = 315°$ in the standard rotation matrix:

$$\begin{bmatrix} \cos\theta & -\sin\theta \\ \sin\theta & \cos\theta \end{bmatrix} = \begin{bmatrix} \cos 315° & -\sin 315° \\ \sin 315° & \cos 315° \end{bmatrix}$$

$$= \begin{bmatrix} \frac{1}{\sqrt{2}} & \frac{1}{\sqrt{2}} \\ -\frac{1}{\sqrt{2}} & \frac{1}{\sqrt{2}} \end{bmatrix} = \frac{1}{\sqrt{2}}\begin{bmatrix} 1 & 1 \\ -1 & 1 \end{bmatrix}$$

A rotation of 45° clockwise is equivalent to a rotation of 315° anticlockwise and $\sin 315° = -\frac{1}{\sqrt{2}}$.

An alternative method here is to simply substitute $\theta = -45°$ into the standard rotation matrix as a rotation of 45° clockwise is equivalent to −45° anticlockwise.

Worked example 6.7

Write down the transformations that have the following matrices:

(a) $\begin{bmatrix} 0 & -1 \\ 1 & 0 \end{bmatrix}$, **(b)** $\begin{bmatrix} \frac{1}{2} & -\frac{\sqrt{3}}{2} \\ \frac{\sqrt{3}}{2} & \frac{1}{2} \end{bmatrix}$,

(c) $\begin{bmatrix} -\frac{\sqrt{3}}{2} & -\frac{1}{2} \\ \frac{1}{2} & -\frac{\sqrt{3}}{2} \end{bmatrix}$, **(d)** $\begin{bmatrix} 0.34 & 0.94 \\ -0.94 & 0.34 \end{bmatrix}$.

Solution

The first step in this sort of question is to recognise that the above matrices represent rotations.

A closer inspection of the standard rotation matrix shows you that the elements of each diagonal have the **same absolute value** but the **signs of one diagonal pair are different**.

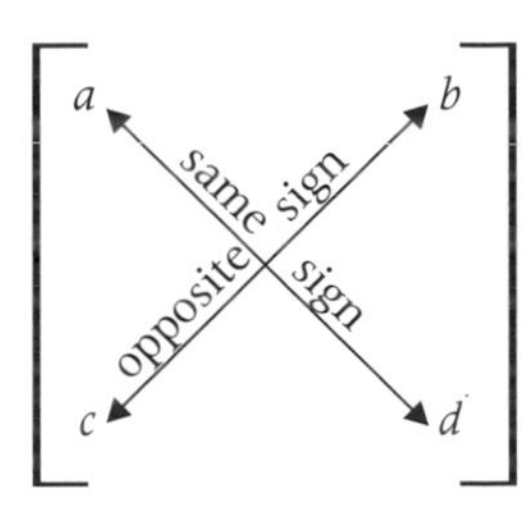

for a rotation matrix.

(a) Comparing $\begin{bmatrix} 0 & -1 \\ 1 & 0 \end{bmatrix}$ with the standard rotation matrix

$\begin{bmatrix} \cos\theta & -\sin\theta \\ \sin\theta & \cos\theta \end{bmatrix}$ gives $\cos\theta = 0$ and $\sin\theta = 1$.

This occurs when $\theta = 90°$.

The matrix represents an anticlockwise rotation of 90° about the origin.

(b) Comparing $\begin{bmatrix} \frac{1}{2} & -\frac{\sqrt{3}}{2} \\ \frac{\sqrt{3}}{2} & \frac{1}{2} \end{bmatrix}$, with the standard rotation matrix

$\begin{bmatrix} \cos\theta & -\sin\theta \\ \sin\theta & \cos\theta \end{bmatrix}$ gives $\cos\theta = \frac{1}{2}$ and $\sin\theta = \frac{\sqrt{3}}{2}$.

θ is therefore in the **first quadrant**.

This occurs when $\theta = \cos^{-1}\left(\frac{1}{2}\right) = 60°$.

The matrix represents an anticlockwise rotation of 60° about the origin.

(c) Comparing $\begin{bmatrix} -\frac{\sqrt{3}}{2} & -\frac{1}{2} \\ \frac{1}{2} & -\frac{\sqrt{3}}{2} \end{bmatrix}$ with the standard rotation matrix

$\begin{bmatrix} \cos\theta & -\sin\theta \\ \sin\theta & \cos\theta \end{bmatrix}$ gives $\cos\theta = -\frac{\sqrt{3}}{2}$ and $\sin\theta = \frac{1}{2}$.

θ is therefore in the **second quadrant**.

This occurs when $\theta = \cos^{-1}\left(-\frac{\sqrt{3}}{2}\right) = 150°$.

The matrix represents an anticlockwise rotation of 150° about the origin.

(d) Comparing $\begin{bmatrix} 0.34 & 0.94 \\ -0.94 & 0.34 \end{bmatrix}$ with the standard rotation matrix

$\begin{bmatrix} \cos\theta & -\sin\theta \\ \sin\theta & \cos\theta \end{bmatrix}$ gives $\cos\theta = 0.34$ and $\sin\theta = -0.94$.

θ is therefore in the **fourth quadrant**.

This occurs when $\theta = \cos^{-1} 0.34 = 289.9°$ (3 s.f.).

The matrix represents an anticlockwise rotation of 289.9° about the origin.

Perhaps a better way of expressing this rotation is as a clockwise rotation of 70.1° about the origin.

6.6 Reflections in a line through the origin

Suppose you want to find the matrix associated with a **reflection** in a line through the origin which makes **angle θ with the positive x-axis**.

Again, to find this matrix you can consider the effect of the reflection on the two points (1, 0) and (0, 1).

Find the image of (1, 0):

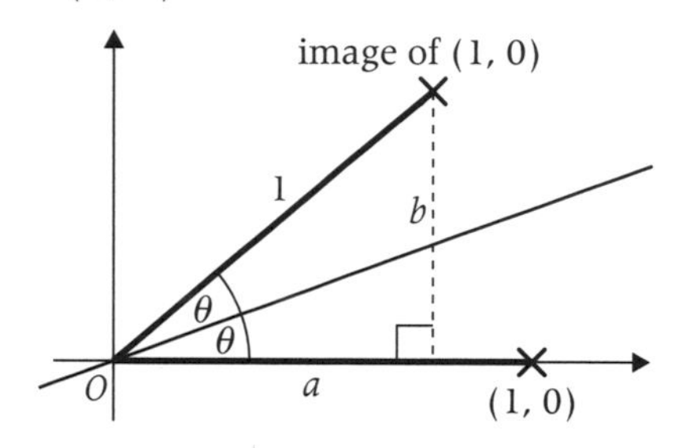

It should be clear from the right-angled triangle that

$$\cos 2\theta = \frac{a}{1} \quad \Rightarrow \quad a = \cos 2\theta$$

and $\sin 2\theta = \dfrac{b}{1} \quad \Rightarrow \quad b = \sin 2\theta.$

So the image of (1, 0) under the reflection is $(\cos 2\theta, \sin 2\theta)$.

Written in column vector form (matrix form), this is

$$\begin{bmatrix} 1 \\ 0 \end{bmatrix} \rightarrow \begin{bmatrix} \cos 2\theta \\ \sin 2\theta \end{bmatrix}.$$

Find the image of (0, 1):

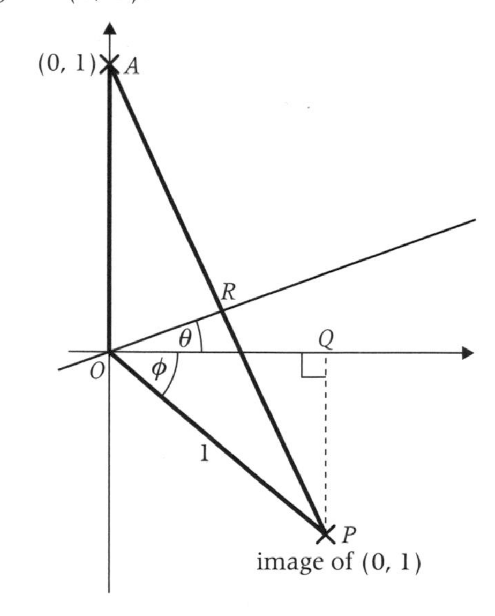

Now angle $OAP = \theta$

$\Rightarrow$ angle $OPA = \theta$, since triangle OAP is isosceles.

Considering triangle ORP, $\phi = 90° - 2\theta$:

from the right-angled triangle OPQ, it should be clear that

$$\cos\phi = \frac{OQ}{1} \Rightarrow OQ = \cos\phi \Rightarrow OQ = \cos(90° - 2\theta)$$

and $$\sin\phi = \frac{PQ}{1} \Rightarrow PQ = \sin\phi \Rightarrow PQ = \sin(90° - 2\theta);$$

but $\cos(90° - 2\theta) = \sin 2\theta$ and $\sin(90° - 2\theta) = \cos 2\theta$

$\Rightarrow$ $OQ = \sin 2\theta$ and $PQ = \cos 2\theta$.

Negative since the y-coordinate of image is negative.

So the image of (0, 1) under the reflection is $(\sin 2\theta, -\cos 2\theta)$. Written in column vector form (matrix form), this is

$$\begin{bmatrix} 0 \\ 1 \end{bmatrix} \rightarrow \begin{bmatrix} \sin 2\theta \\ -\cos 2\theta \end{bmatrix}.$$

Putting the two results together gives the reflection matrix as

$$\begin{bmatrix} \cos 2\theta & \sin 2\theta \\ \sin 2\theta & -\cos 2\theta \end{bmatrix}.$$

↑ Image of (1, 0). ↑ Image of (0, 1).

It is also necessary to realise that the straight line that makes an angle of θ with the positive x-axis and passes through the origin has the equation $y = x\tan\theta$.

So,

A reflection in the line $y = x\tan\theta$ (where θ is the angle the line makes with the positive x-axis) is given by:

$$\begin{bmatrix} x' \\ y' \end{bmatrix} = \begin{bmatrix} \cos 2\theta & \sin 2\theta \\ \sin 2\theta & -\cos 2\theta \end{bmatrix}\begin{bmatrix} x \\ y \end{bmatrix}.$$

This matrix is in the formulae booklet provided in the examination.

You should notice that the standard reflection matrix is very similar to the standard rotation matrix but there are enough differences for you to spot.

Again, a closer inspection of the signs and values of the elements of the diagonals does help.

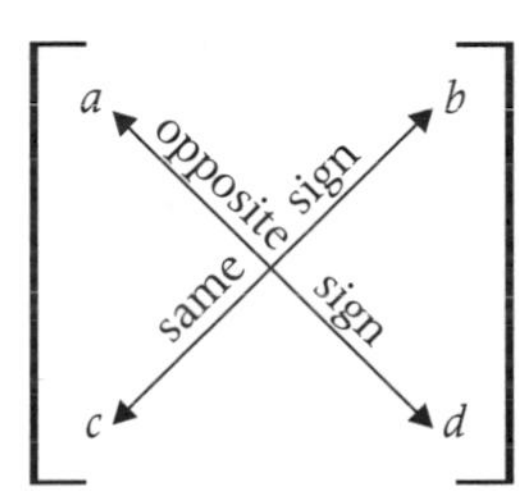

for a reflection matrix.

Worked example 6.8

Give a full description of the transformations represented by the following matrices.

(a) $\begin{bmatrix} -1 & 0 \\ 0 & 1 \end{bmatrix}$, (b) $\begin{bmatrix} \frac{1}{2} & \frac{\sqrt{3}}{2} \\ \frac{\sqrt{3}}{2} & -\frac{1}{2} \end{bmatrix}$, (c) $\frac{1}{\sqrt{5}}\begin{bmatrix} 1 & 2 \\ 2 & -1 \end{bmatrix}$.

Solution

Again, by considering the signs and absolute values of the elements forming the diagonals you should recognise all of the above matrices as representing reflections.

(a) Comparing $\begin{bmatrix} -1 & 0 \\ 0 & 1 \end{bmatrix}$ with the standard reflection matrix

$\begin{bmatrix} \cos 2\theta & \sin 2\theta \\ \sin 2\theta & -\cos 2\theta \end{bmatrix}$ gives $\cos 2\theta = -1$ and $\sin 2\theta = 0$.

This occurs when $2\theta = 180° \Rightarrow \theta = 90°$.

The matrix represents a reflection in the line which makes an angle of 90° with the positive x-axis, i.e. this is a reflection in the y-axis.

(b) Comparing $\begin{bmatrix} \frac{1}{2} & \frac{\sqrt{3}}{2} \\ \frac{\sqrt{3}}{2} & -\frac{1}{2} \end{bmatrix}$ with the standard reflection matrix

$\begin{bmatrix} \cos 2\theta & \sin 2\theta \\ \sin 2\theta & -\cos 2\theta \end{bmatrix}$ gives $\cos 2\theta = \frac{1}{2}$ and $\sin 2\theta = \frac{\sqrt{3}}{2}$.

2θ is in the **first quadrant.**

This occurs when $2\theta = \cos^{-1}\left(\frac{1}{2}\right) = 60° \Rightarrow \theta = 30°$.

The matrix represents a reflection in the line which makes an angle of 30° with the positive x-axis, i.e. this is a reflection in the line $y = x \tan 30°$ or $y = \frac{x}{\sqrt{3}}$.

(c) Comparing $\begin{bmatrix} \frac{1}{\sqrt{5}} & \frac{2}{\sqrt{5}} \\ \frac{2}{\sqrt{5}} & -\frac{1}{\sqrt{5}} \end{bmatrix}$ with the standard reflection matrix

$\begin{bmatrix} \cos 2\theta & \sin 2\theta \\ \sin 2\theta & -\cos 2\theta \end{bmatrix}$ gives $\cos 2\theta = \frac{1}{\sqrt{5}}$ and $\sin 2\theta = \frac{2}{\sqrt{5}}$.

2θ is in the **first quadrant.**

This occurs when $2\theta = \cos^{-1}\left(\frac{1}{\sqrt{5}}\right) = 63.4°$ (3 s.f.)

$\Rightarrow \theta = 31.7°$ (3 s.f.).

The matrix represents a reflection in the line $y = x \tan \theta$ where $\theta = 31.7°$ (3 s.f.).

Worked example 6.9

Find the transformation matrix that represents:

(a) a reflection in the x-axis,

(b) a reflection in the line $y = -x$.

Solution

(a) The x-axis makes an angle of 0° with the positive x-axis. Substituting $\theta = 0°$ in the standard reflection matrix gives:

$$\begin{bmatrix} \cos 2\theta & \sin 2\theta \\ \sin 2\theta & -\cos 2\theta \end{bmatrix} = \begin{bmatrix} \cos 0° & \sin 0° \\ \sin 0° & -\cos 0° \end{bmatrix} = \begin{bmatrix} 1 & 0 \\ 0 & -1 \end{bmatrix}.$$

(b) The line $y = -x$ makes an angle of 135° with the positive x-axis. Substituting $\theta = 135°$ in the standard reflection matrix gives:

$$\begin{bmatrix} \cos 2\theta & \sin 2\theta \\ \sin 2\theta & -\cos 2\theta \end{bmatrix} = \begin{bmatrix} \cos 270° & \sin 270° \\ \sin 270° & -\cos 270° \end{bmatrix} = \begin{bmatrix} 0 & -1 \\ -1 & 0 \end{bmatrix}.$$

Worked example 6.10

Find the matrix that represents a reflection in the line $y = 3x$.

6

Solution

The reflection line is $y = 3x \quad \Rightarrow \quad \tan\theta = 3$.

In order to write down the matrix, you need the values of $\cos 2\theta$ and $\sin 2\theta$.

Since you know that $\tan\theta = 3$, you can draw a right-angled triangle:

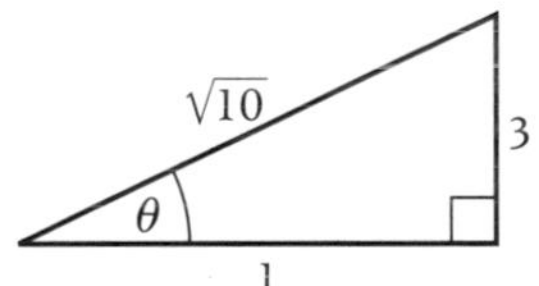

By Pythagoras, the hypotenuse is $\sqrt{10}$.

From the diagram, it should be clear that $\sin\theta = \dfrac{3}{\sqrt{10}}$ and $\cos\theta = \dfrac{1}{\sqrt{10}}$.

To find the values of $\cos 2\theta$ and $\sin 2\theta$ you can use the double angle identities:

$$\sin 2\theta = 2\sin\theta\cos\theta = 2 \times \frac{3}{\sqrt{10}} \times \frac{1}{\sqrt{10}} = \frac{6}{10} = \frac{3}{5}$$

and $$\cos 2\theta = 2\cos^2\theta - 1 = 2 \times \left(\frac{1}{\sqrt{10}}\right)^2 - 1 = \frac{2}{10} - 1 = -\frac{4}{5}.$$

Substituting these values into the standard reflection matrix gives:

$$\begin{bmatrix} \cos 2\theta & \sin 2\theta \\ \sin 2\theta & -\cos 2\theta \end{bmatrix} = \begin{bmatrix} -\frac{4}{5} & \frac{3}{5} \\ \frac{3}{5} & \frac{4}{5} \end{bmatrix}.$$

EXERCISE 6B

1 Give a full description of the transformations given by the following matrices:

(a) $\begin{bmatrix} -1 & 0 \\ 0 & -1 \end{bmatrix}$, **(b)** $\begin{bmatrix} -\frac{1}{\sqrt{2}} & -\frac{1}{\sqrt{2}} \\ \frac{1}{\sqrt{2}} & -\frac{1}{\sqrt{2}} \end{bmatrix}$,

(c) $\begin{bmatrix} -\frac{1}{2} & \frac{\sqrt{3}}{2} \\ \frac{\sqrt{3}}{2} & \frac{1}{2} \end{bmatrix}$, **(d)** $\begin{bmatrix} \frac{1}{\sqrt{2}} & \frac{1}{\sqrt{2}} \\ \frac{1}{\sqrt{2}} & -\frac{1}{\sqrt{2}} \end{bmatrix}$,

(e) $\begin{bmatrix} 0 & 1 \\ 1 & 0 \end{bmatrix}$, **(f)** $\begin{bmatrix} 0.766 & 0.643 \\ -0.643 & 0.766 \end{bmatrix}$,

(g) $\begin{bmatrix} 0.342 & 0.940 \\ 0.940 & -0.342 \end{bmatrix}$, **(h)** $\begin{bmatrix} \frac{2}{\sqrt{5}} & -\frac{1}{\sqrt{5}} \\ \frac{1}{\sqrt{5}} & \frac{2}{\sqrt{5}} \end{bmatrix}$,

(i) $\begin{bmatrix} \frac{2}{\sqrt{5}} & \frac{1}{\sqrt{5}} \\ \frac{1}{\sqrt{5}} & -\frac{2}{\sqrt{5}} \end{bmatrix}$, **(j)** $\begin{bmatrix} -\frac{1}{9} & \frac{4\sqrt{5}}{9} \\ \frac{4\sqrt{5}}{9} & \frac{1}{9} \end{bmatrix}$,

(k) $\begin{bmatrix} 0.6 & 0.8 \\ -0.8 & 0.6 \end{bmatrix}$, **(l)** $\begin{bmatrix} -0.28 & -0.96 \\ -0.96 & 0.28 \end{bmatrix}$.

2 Write down the matrices that are associated with the following transformations:

(a) a reflection in the line through the origin which makes an angle of 30° with the positive x-axis,

(b) an anticlockwise rotation of 300° about the origin,

(c) a clockwise rotation of 45° about the origin,

(d) a reflection in the line through the origin which makes an angle of 60° with the positive x-axis,

(e) a rotation through 135° anticlockwise about the origin,

(f) a rotation through $\cos^{-1}\left(\frac{1}{3}\right)$, clockwise about the origin,

(g) a reflection in the line $y = 4x$,

(h) a reflection in the line $y = -3x$.

3 A transformation is given by $x' = -\frac{x}{2} - \frac{\sqrt{3}y}{2}$, $y' = -\frac{\sqrt{3}x}{2} + \frac{y}{2}$.

(a) Find the exact coordinates of the image of the point $P(4, 2)$ under this transformation.

(b) Give a full geometrical description of the transformation.

4 A transformation is given by $x' = \frac{1}{\sqrt{2}}(x - y)$, $y' = \frac{1}{\sqrt{2}}(x + y)$

(a) Write down the matrix that represents this transformation.

(b) Under this transformation, the point A has image $(2, 2)$. Find the exact coordinates of A.

(c) Give a full description of the transformation.

6.7 Composite transformations

When you perform one transformation of the plane followed by another the overall result is also a transformation.

A transformation formed by applying successive transformations is referred to as a **composite transformation**.

In exactly the same manner as you form composite functions.

For example, if you perform an anticlockwise rotation through 90° about the origin followed by a reflection in the line $y = x$, then the overall effect is the same as performing one reflection in the x-axis.

You can prove that this is true by finding the matrix that represents the composite transformation. Worked example 6.11 shows that this can be done by finding the images of the points (1, 0) and (0, 1) in the same manner as you did when finding the standard rotation and reflection matrices.

Worked example 6.11

Prove that an anticlockwise rotation through 90° about the origin followed by a reflection in the line $y = x$ is the same as performing a reflection in the x-axis.

Solution

The matrix that represents the rotation is $\mathbf{A} = \begin{bmatrix} 0 & -1 \\ 1 & 0 \end{bmatrix}$.

The matrix that represents the reflection is $\mathbf{B} = \begin{bmatrix} 0 & 1 \\ 1 & 0 \end{bmatrix}$.

To find the image of (1, 0), first perform the rotation

$$\begin{bmatrix} 0 & -1 \\ 1 & 0 \end{bmatrix}\begin{bmatrix} 1 \\ 0 \end{bmatrix} = \begin{bmatrix} 0 \\ 1 \end{bmatrix}$$

then the reflection $\begin{bmatrix} 0 & 1 \\ 1 & 0 \end{bmatrix}\begin{bmatrix} 0 \\ 1 \end{bmatrix} = \begin{bmatrix} 1 \\ 0 \end{bmatrix}$.

So the image of (1, 0) after the two transformations is (1, 0). Written in column vector form (matrix form), this is

$$\begin{bmatrix} 1 \\ 0 \end{bmatrix} \rightarrow \begin{bmatrix} 1 \\ 0 \end{bmatrix}.$$

To find the image of (0, 1), first perform the rotation

$$\begin{bmatrix} 0 & -1 \\ 1 & 0 \end{bmatrix}\begin{bmatrix} 0 \\ 1 \end{bmatrix} = \begin{bmatrix} -1 \\ 0 \end{bmatrix}$$

Then the reflection $\begin{bmatrix} 0 & 1 \\ 1 & 0 \end{bmatrix}\begin{bmatrix} -1 \\ 0 \end{bmatrix} = \begin{bmatrix} 0 \\ -1 \end{bmatrix}$.

So the image of (0, 1) after the two transformations is (0, −1). Written in column vector form (matrix form), this is

$$\begin{bmatrix} 0 \\ 1 \end{bmatrix} \rightarrow \begin{bmatrix} 0 \\ -1 \end{bmatrix}$$

Putting the two results together gives the matrix

$$\begin{bmatrix} 1 & 0 \\ 0 & -1 \end{bmatrix}$$

↑ Image of (1, 0). ↑ Image of (0, 1).

This matrix represents a reflection in the x-axis.

You can see from the previous example how the two matrices representing the transformations are applied. In fact, rather than finding the images of the points (1, 0) and (0, 1) you can find the matrix representing the composite much quicker by simply multiplying the two matrices together **but** noticing that you always have to **pre-multiply** by a transformation matrix.

To find the image of a point (x, y) after the successive transformations represented by $\mathbf{A} = \begin{bmatrix} a & b \\ c & d \end{bmatrix}$ followed by $\mathbf{B} = \begin{bmatrix} e & f \\ g & h \end{bmatrix}$ you have to first **pre-multiply** by **A** then **pre-multiply** the result by **B**, i.e.,

$$\begin{bmatrix} x' \\ y' \end{bmatrix} = \begin{bmatrix} e & f \\ g & h \end{bmatrix} \begin{bmatrix} a & b \\ c & d \end{bmatrix} \begin{bmatrix} x \\ y \end{bmatrix}.$$

The overall result is given by the product matrix **BA**.

The order of multiplication is absolutely vital here since matrix multiplication is not commutative, i.e. the products **BA** and **AB** are different and only **BA** gives the correct result.

In summary,

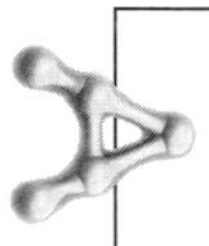

If the 2×2 matrices, **A** and **B**, both represent transformations, then the product matrix **BA** also represents a transformation equivalent to applying **A** followed by **B**.

This type of transformation is known as a composite transformation.

It is important to remember to read composite transformations backwards, i.e.,

ABC means **C** first then **B** then **A**.

Worked example 6.12

The transformation **A** is a reflection in the line $y = \frac{x}{\sqrt{3}}$ and the transformation **B** is a reflection in the line $y = \sqrt{3}x$.

The composite transformation **T** is formed by applying **A** followed by **B**.

Give a full geometrical description of **T**.

Solution

First you need to find the matrices that represent transformations **A** and **B**.

A the line $y = \frac{x}{\sqrt{3}}$ makes an angle of $\theta = \tan^{-1}\left(\frac{1}{\sqrt{3}}\right) = 30°$ with the positive x-axis.

The matrix is $\mathbf{A} = \begin{bmatrix} \cos 60° & \sin 60° \\ \sin 60° & -\cos 60° \end{bmatrix} = \begin{bmatrix} \frac{1}{2} & \frac{\sqrt{3}}{2} \\ \frac{\sqrt{3}}{2} & -\frac{1}{2} \end{bmatrix}$

$$= \frac{1}{2}\begin{bmatrix} 1 & \sqrt{3} \\ \sqrt{3} & -1 \end{bmatrix}$$

B the line $y = \sqrt{3}x$ makes an angle of $\theta = \tan^{-1}(\sqrt{3}) = 60°$ with the positive x-axis.

The matrix $\mathbf{B} = \begin{bmatrix} \cos 120° & \sin 120° \\ \sin 120° & -\cos 120° \end{bmatrix} = \begin{bmatrix} -\frac{1}{2} & \frac{\sqrt{3}}{2} \\ \frac{\sqrt{3}}{2} & \frac{1}{2} \end{bmatrix}$

$$= \frac{1}{2}\begin{bmatrix} -1 & \sqrt{3} \\ \sqrt{3} & 1 \end{bmatrix}$$

So $\mathbf{T} = \mathbf{BA} = \frac{1}{2}\begin{bmatrix} -1 & \sqrt{3} \\ \sqrt{3} & 1 \end{bmatrix} \times \frac{1}{2}\begin{bmatrix} 1 & \sqrt{3} \\ \sqrt{3} & -1 \end{bmatrix}$

$$= \frac{1}{4}\begin{bmatrix} -1 & \sqrt{3} \\ \sqrt{3} & 1 \end{bmatrix} \begin{bmatrix} 1 & \sqrt{3} \\ \sqrt{3} & -1 \end{bmatrix}$$

$$= \frac{1}{4}\begin{bmatrix} 2 & -2\sqrt{3} \\ 2\sqrt{3} & 2 \end{bmatrix}$$

$$= \begin{bmatrix} \frac{1}{2} & -\frac{\sqrt{3}}{2} \\ \frac{\sqrt{3}}{2} & \frac{1}{2} \end{bmatrix}$$

You should recognise this as a rotation matrix.

$$= \begin{bmatrix} \cos 60° & -\sin 60° \\ \sin 60° & \cos 60° \end{bmatrix}$$

T is a rotation of 60° anticlockwise about the origin.

Worked example 6.13

A, **B** and **C** are the transformations:

A a rotation of 30° anticlockwise about the origin;

B a reflection in the line $y = \frac{x}{\sqrt{3}}$;

C a rotation of 150° anticlockwise about the origin.

Transformation **T** is formed by applying **C** then **B** then **A**. Describe fully the composite transformation **T**.

Solution

First you need to find the three transformation matrices.

A $\begin{bmatrix} \cos 30° & -\sin 30° \\ \sin 30° & \cos 30° \end{bmatrix} = \begin{bmatrix} \frac{\sqrt{3}}{2} & -\frac{1}{2} \\ \frac{1}{2} & \frac{\sqrt{3}}{2} \end{bmatrix} = \frac{1}{2}\begin{bmatrix} \sqrt{3} & -1 \\ 1 & \sqrt{3} \end{bmatrix}$

B The line $y = \frac{x}{\sqrt{3}}$ makes an angle of $\theta = \tan^{-1}\left(\frac{1}{\sqrt{3}}\right) = 30°$ with the positive x-axis.

$\begin{bmatrix} \cos 60° & \sin 60° \\ \sin 60° & -\cos 60° \end{bmatrix} = \begin{bmatrix} \frac{1}{2} & \frac{\sqrt{3}}{2} \\ \frac{\sqrt{3}}{2} & -\frac{1}{2} \end{bmatrix} = \frac{1}{2}\begin{bmatrix} 1 & \sqrt{3} \\ \sqrt{3} & -1 \end{bmatrix}$

C $\begin{bmatrix} \cos 150° & -\sin 150° \\ \sin 150° & \cos 150° \end{bmatrix} = \begin{bmatrix} -\frac{\sqrt{3}}{2} & -\frac{1}{2} \\ \frac{1}{2} & -\frac{\sqrt{3}}{2} \end{bmatrix} = \frac{1}{2}\begin{bmatrix} -\sqrt{3} & -1 \\ 1 & -\sqrt{3} \end{bmatrix}$

$$\mathbf{T} = \mathbf{ABC} = \frac{1}{2}\begin{bmatrix} \sqrt{3} & -1 \\ 1 & \sqrt{3} \end{bmatrix} \times \frac{1}{2}\begin{bmatrix} 1 & \sqrt{3} \\ \sqrt{3} & -1 \end{bmatrix} \times \frac{1}{2}\begin{bmatrix} -\sqrt{3} & -1 \\ 1 & -\sqrt{3} \end{bmatrix}$$

$$= \frac{1}{8}\begin{bmatrix} \sqrt{3} & -1 \\ 1 & \sqrt{3} \end{bmatrix} \begin{bmatrix} 1 & \sqrt{3} \\ \sqrt{3} & -1 \end{bmatrix} \begin{bmatrix} -\sqrt{3} & -1 \\ 1 & -\sqrt{3} \end{bmatrix}$$

$$= \frac{1}{8}\begin{bmatrix} 0 & 4 \\ 4 & 0 \end{bmatrix} \begin{bmatrix} -\sqrt{3} & -1 \\ 1 & -\sqrt{3} \end{bmatrix}$$

$$= \frac{1}{8}\begin{bmatrix} 4 & -4\sqrt{3} \\ -4\sqrt{3} & -4 \end{bmatrix}$$

$$= \begin{bmatrix} \frac{1}{2} & -\frac{\sqrt{3}}{2} \\ -\frac{\sqrt{3}}{2} & -\frac{1}{2} \end{bmatrix}$$

$$= \begin{bmatrix} \cos 300° & \sin 300° \\ \sin 300° & -\cos 300° \end{bmatrix}$$

So **T** is a reflection in the line $y = x \tan 150°$, i.e. a reflection in the line $y = -\frac{x}{\sqrt{3}}$.

Worked example 6.14

(a) Give a full geometrical description of the plane transformation represented by the matrix **A**, where $\mathbf{A} = \begin{bmatrix} 3 & -\sqrt{7} \\ \sqrt{7} & 3 \end{bmatrix}$

(b) Give a full geometric description of the single transformation represented by the matrix $\mathbf{A}^2$.

Solution

(a) At first glance the matrix **A** seems to be a rotation matrix. On closer inspection, however, there is a problem as comparing **A** with the standard rotation matrix gives $\cos\theta = 3$ and $\sin\theta = \sqrt{7}$, which is impossible since the maximum value of both sine and cosine is 1.

There is, fortunately, a little 'trick' that gets us around this problem.

Assume that $\cos\theta = \frac{3}{h}$ and $\sin\theta = \frac{\sqrt{7}}{h}$, and write **A** as

$$\mathbf{A} = \begin{bmatrix} h & 0 \\ 0 & h \end{bmatrix} \begin{bmatrix} \frac{3}{h} & -\frac{\sqrt{7}}{h} \\ \frac{\sqrt{7}}{h} & \frac{3}{h} \end{bmatrix}.$$

Multiply out the matrices to convince yourself that it is the same as **A**.

Now you need to find the value of h.

To do this you draw a right-angled triangle using the data you know.

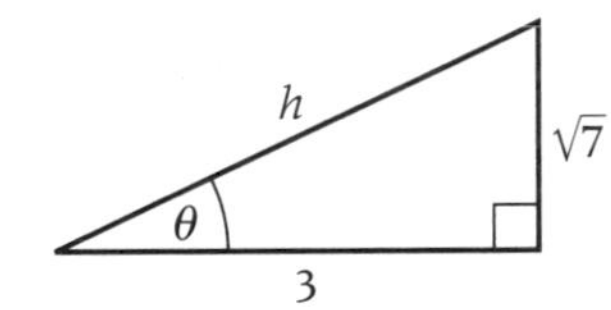

By Pythagoras, $h = \sqrt{9 + 7} = 4$.

So $\mathbf{A} = \begin{bmatrix} 4 & 0 \\ 0 & 4 \end{bmatrix} \begin{bmatrix} \frac{3}{4} & -\frac{\sqrt{7}}{4} \\ \frac{\sqrt{7}}{4} & \frac{3}{4} \end{bmatrix}.$

Now $\begin{bmatrix} \frac{3}{4} & -\frac{\sqrt{7}}{4} \\ \frac{\sqrt{7}}{4} & \frac{3}{4} \end{bmatrix}$ is an anticlockwise rotation of

$\cos^{-1}\left(\frac{3}{4}\right) = 41.41°$ about the origin and $\begin{bmatrix} 4 & 0 \\ 0 & 4 \end{bmatrix}$ is an enlargement with scale factor 4, centre the origin.

A is an anticlockwise rotation of 41.41° about the origin followed by an enlargement of scale factor 4, centre the origin.

(b) $\mathbf{A}^2 = \begin{bmatrix} 3 & -\sqrt{7} \\ \sqrt{7} & 3 \end{bmatrix} \begin{bmatrix} 3 & -\sqrt{7} \\ \sqrt{7} & 3 \end{bmatrix} = \begin{bmatrix} 2 & -6\sqrt{7} \\ 6\sqrt{7} & 2 \end{bmatrix}$

Using the same procedure as in **(a)** gives

$$\mathbf{A}^2 = \begin{bmatrix} 16 & 0 \\ 0 & 16 \end{bmatrix} \begin{bmatrix} \frac{2}{16} & -\frac{6\sqrt{7}}{16} \\ \frac{6\sqrt{7}}{16} & \frac{2}{16} \end{bmatrix}.$$

Which represents a rotation of 82.82° anticlockwise about the origin followed by an enlargement, centre the origin and scale factor 16.

EXERCISE 6C

1 The composite transformation **T** is formed by applying a rotation of 30° anticlockwise about the origin followed by a stretch in the x-direction of factor 2. Find the matrix that represents **T**.

2 The composite transformation **C** is formed by first reflecting in the line $y = x$ followed by an enlargement with centre the origin and scale factor 5. Find the matrix that represents **C**.

3 The composite transformation **M** is formed by applying a rotation of 90° anticlockwise about the origin followed by a reflection in the line $y = -x$.

(a) Find the matrix that represents **M**.

(b) Give a full description of transformation **M**.

4 The matrix **A** represents a rotation of 60° anticlockwise about the origin.

(a) Find matrix **A**.

(b) Give a full geometric description of the transformation represented by the matrix $\mathbf{A}^2$.

5 The transformation **A** is a reflection in the line $y = x$ and the transformation **B** is a reflection in the line $y = -x$. The composite transformation **T** is formed by applying **B** followed by **A**. Give a complete description of **T**.

6 Prove that a composite transformation formed by two successive reflections in any two straight lines through the origin that are perpendicular is equivalent to performing a half-turn about the origin.

7 The composite transformation **M** is defined by **M** = **ABC** where **A**, **B** and **C** are the transformations:

A a reflection in the line through the origin at 30° to the positive x-axis;

B a stretch in the y-direction of factor 2;

C a rotation of 45° anticlockwise about the origin.

Find the matrix that represents **M**.

8 **A**, **B** and **C** are the transformations:

A a rotation of 30° anticlockwise about the origin;

B a reflection in the line $y = \sqrt{3}x$;

C a rotation of 150° anticlockwise about the origin.

Transformation **T** is formed by applying **C** then **B** then **A**. Describe fully the composite transformation **T**.

9 Give a full geometrical description of the transformation represented by the matrix $\begin{bmatrix} 3 & -4 \\ 4 & 3 \end{bmatrix}$.

MIXED EXERCISE

1 A transformation **T** is given by $\begin{bmatrix} x' \\ y' \end{bmatrix} = \frac{1}{5}\begin{bmatrix} 3 & 4 \\ -4 & 3 \end{bmatrix}\begin{bmatrix} x \\ y \end{bmatrix}$.

(a) Find the image of each of the points $A(5, 0)$ and $B(0, 5)$.

(b) Describe fully the transformation represented by **T**. [A]

2 The transformation **T** has matrix $\mathbf{A} = \begin{bmatrix} \frac{3}{5} & \frac{4}{5} \\ \frac{4}{5} & -\frac{3}{5} \end{bmatrix}$.

Give a full geometrical description of **T**. [A]

3 (a) Given that $\tan\theta = \frac{1}{\sqrt{5}}$, prove that $\sin 2\theta = \frac{\sqrt{5}}{3}$ and find the value of $\cos 2\theta$.

(b) Give a full geometrical description of the plane transformation represented by the matrix **M**, where

$$\mathbf{M} = \begin{bmatrix} 2 & \sqrt{5} \\ \sqrt{5} & -2 \end{bmatrix}.$$

(c) Give a full geometrical description of the single transformation represented by the matrix $\mathbf{M}^2$. [A]

4 Let **M** be the matrix $\begin{bmatrix} 2 & -1 \\ -2 & 3 \end{bmatrix}$.

A transformation of the plane **T** is such that

$$\begin{bmatrix} x' \\ y' \end{bmatrix} = \mathbf{M}\begin{bmatrix} x \\ y \end{bmatrix}.$$

Find the point which is mapped onto the point (8, 4) under **T**. [A]

Key point summary

1 A linear transformation that changes/transforms point $P(x, y)$ into point $P'(x', y')$ can be written as: *p63*

$$\begin{bmatrix} x' \\ y' \end{bmatrix} = \begin{bmatrix} a & b \\ c & d \end{bmatrix}\begin{bmatrix} x \\ y \end{bmatrix} \quad \textbf{matrix form}$$

or $x' = ax + by$ **algebraic form**

$y' = cx + dy$

2 The matrix $\begin{bmatrix} a & b \\ c & d \end{bmatrix}$ is called a transformation matrix. *p63*

3 A one-way stretch in the x-direction of scale factor k is given by: *p66*

$$\begin{bmatrix} x' \\ y' \end{bmatrix} = \begin{bmatrix} k & 0 \\ 0 & 1 \end{bmatrix}\begin{bmatrix} x \\ y \end{bmatrix}.$$

6

4 A one-way stretch in the y-direction of scale factor k is given by: *p66*

$$\begin{bmatrix} x' \\ y' \end{bmatrix} = \begin{bmatrix} 1 & 0 \\ 0 & k \end{bmatrix} \begin{bmatrix} x \\ y \end{bmatrix}.$$

5 A two-way stretch of scale factor a in the x-direction and scale factor b in the y-direction is given by: *p66*

$$\begin{bmatrix} x' \\ y' \end{bmatrix} = \begin{bmatrix} a & 0 \\ 0 & b \end{bmatrix} \begin{bmatrix} x \\ y \end{bmatrix}.$$

6 An enlargement of scale factor k with centre the origin is given by: *p67*

$$\begin{bmatrix} x' \\ y' \end{bmatrix} = \begin{bmatrix} k & 0 \\ 0 & k \end{bmatrix} \begin{bmatrix} x \\ y \end{bmatrix}.$$

7 A rotation through angle θ anticlockwise about the origin is given by: *p69*

$$\begin{bmatrix} x' \\ y' \end{bmatrix} = \begin{bmatrix} \cos\theta & -\sin\theta \\ \sin\theta & \cos\theta \end{bmatrix} \begin{bmatrix} x \\ y \end{bmatrix}.$$

This matrix is given in the formulae book.

8 A reflection in the line $y = x \tan\theta$ (where θ is the angle the line makes with the positive x-axis) is given by: *p73*

$$\begin{bmatrix} x' \\ y' \end{bmatrix} = \begin{bmatrix} \cos 2\theta & \sin 2\theta \\ \sin 2\theta & -\cos 2\theta \end{bmatrix} \begin{bmatrix} x \\ y \end{bmatrix}.$$

This matrix is given in the formulae book.

9 If the 2×2 matrices, **A** and **B**, both represent transformations, then the product matrix **BA** also represents a transformation equivalent to applying **A** followed by **B**. *p78*

This type of transformation is known as a composite transformation.

It is important to remember to read composite transformations backwards, i.e.,

ABC means **C** first then **B** then **A**.

Test yourself	What to review
1 A transformation **T** is given by $\begin{bmatrix} x' \\ y' \end{bmatrix} = \begin{bmatrix} 5 & -1 \\ 3 & 2 \end{bmatrix}\begin{bmatrix} x \\ y \end{bmatrix}$. **(a)** Find the image of the point (3, 4) under **T**. **(b)** Find the point which is mapped onto the point (9, 8) under **T**.	*Section 6.2*
2 A transformation is given by $\begin{bmatrix} x' \\ y' \end{bmatrix} = \mathbf{M}\begin{bmatrix} x \\ y \end{bmatrix}$. Describe the geometrical transformation in the cases where: **(a)** $\mathbf{M} = \begin{bmatrix} 2 & 0 \\ 0 & 3 \end{bmatrix}$, **(b)** $\mathbf{M} = \begin{bmatrix} 4 & 0 \\ 0 & 4 \end{bmatrix}$ **(c)** $\mathbf{M} = \begin{bmatrix} 1 & 0 \\ 0 & -1 \end{bmatrix}$.	*Section 6.4*
3 (a) Find the matrix **A** representing a rotation of 30° anticlockwise about the origin. **(b)** Find the matrix **B** representing a reflection in the line $y = x \tan 15°$. **(c)** Hence find the product **AB** and describe the effect of the composite transformation **AB**.	*Sections 6.5 and 6.6*

Test yourself ANSWERS

1 (a) (11, 17); **(b)** (2, 1).

2 (a) Two way stretch of scale factor 2 in x-direction and 3 in y-direction;

(b) Enlargement centre the origin with scale factor 4;

(c) Reflection in x-axis.

3 (a) $\begin{bmatrix} \frac{\sqrt{3}}{2} & -\frac{1}{2} \\ \frac{1}{2} & \frac{\sqrt{3}}{2} \end{bmatrix}$; **(b)** $\begin{bmatrix} \frac{\sqrt{3}}{2} & \frac{1}{2} \\ \frac{1}{2} & -\frac{\sqrt{3}}{2} \end{bmatrix}$; **(c)** $\begin{bmatrix} \frac{1}{2} & \frac{\sqrt{3}}{2} \\ \frac{\sqrt{3}}{2} & -\frac{1}{2} \end{bmatrix}$,

reflection in line $y = x \tan 30°$ or $y = \frac{x}{\sqrt{3}}$.

CHAPTER 7

Linear laws

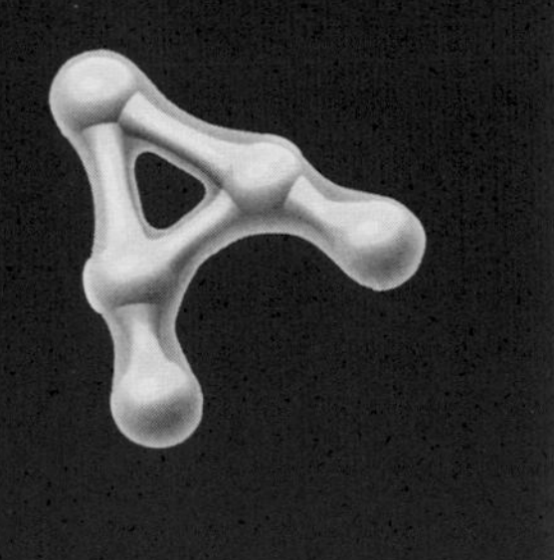

Learning objectives

After studying this chapter, you should be able to:

- use a straight line graph to find unknown constants
- reduce certain kinds of relations between two variables into a linear law
- use logarithms to reduce equations of the form $y = ax^n$ and $y = ab^x$ to a linear law and to use the corresponding linear graphs to find the constants a, n and x.

7.1 A review of straight line graphs and their use in finding unknown constants

Scientists have frequently tried to use experimental data to obtain a relationship between two variables. The easiest way to do this is by means of a straight line graph.

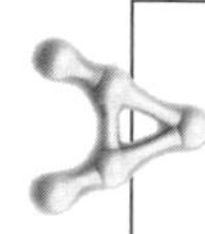

Since experimental data is not exact, due to measuring errors, points plotted are unlikely to all lie exactly in line so a line of best fit is drawn.

In C1 section 3.6 you were shown that

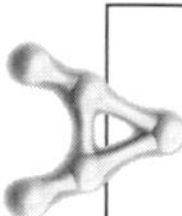

$y = mx + c$ is the equation of a straight line with gradient m and y-intercept c.

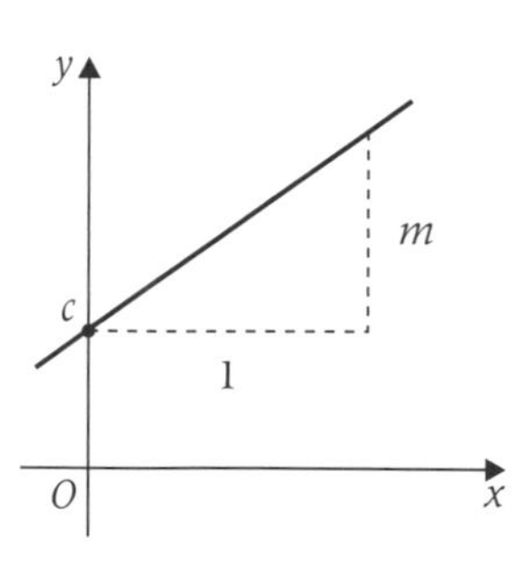

You can find the value of m by taking any two points (x_1, y_1) and (x_2, y_2) **on** the line and using $m = \dfrac{y_2 - y_1}{x_2 - x_1}$.

A common error is to use coordinates of points from a table which may not always lie on the line.

You can usually find the value of c by reading the y-coordinate of the point where the line intersects the y-axis.

Worked example 7.1

The scatter diagram shows the results of a scientific experiment between two variables y and x. A line of best fit has been drawn.

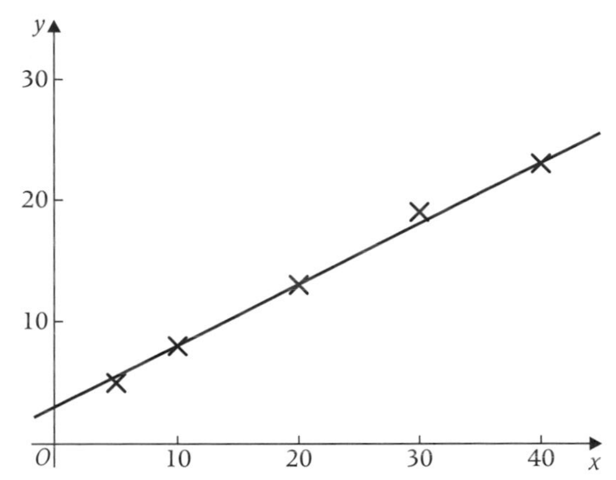

(a) Find the equation of the line of best fit.

(b) Use your equation to estimate the value of y when $x = 80$.

Solution

(a) Points (10, 8) and (40, 23) both lie on the line of best fit.

It is important that you do not use the point (5, 5) or the point (30, 19) as these points do not lie on the drawn line of best fit.

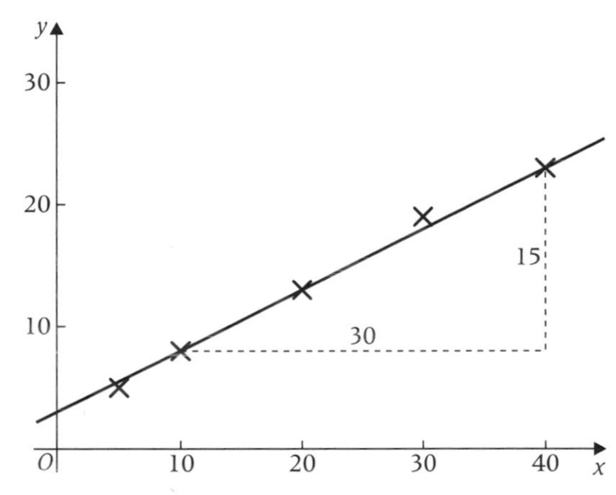

The gradient $m = \dfrac{23 - 8}{40 - 10} = \dfrac{15}{30} = \dfrac{1}{2}$.

The constant $c = y$-intercept $= 3$.

Using the general equation of a line as $y = mx + c$, the equation of the line of best fit is $y = \dfrac{1}{2}x + 3$.

(b) When $x = 80$, $\quad y = \dfrac{1}{2} \times 80 + 3 = 43$.

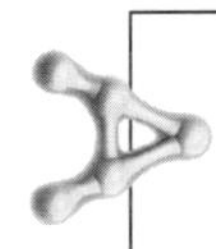

If the variables used are not x and y, the method to find the equation of the line of best fit is exactly the same. In the general equation $y = mx + c$ you just replace y by the variable on the vertical axis and replace x by the variable on the horizontal axis.

7

In the next worked example you will consider a relation between the variables v and t (for example, velocity and time).

Worked example 7.2

The scatter diagram shows the results of a scientific experiment between two variables v and t. A line of best fit has been drawn.

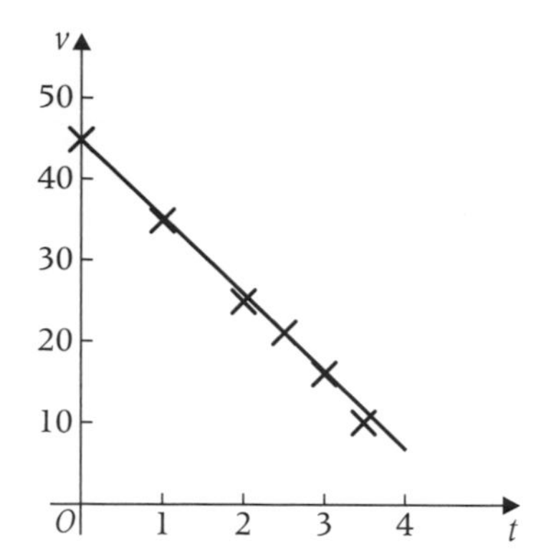

(a) Find the equation of the line of best fit.

(b) Use your equation to estimate the value of t when $v = 0$.

Solution

(a) Variable v is on the vertical axis and variable t is on the horizontal axis, so you write the equation of the line of best fit in the form $v = mt + c$, where m is the gradient and c is the intercept on the v-axis.

Points (0, 45) and (2.5, 20.5) both lie on the line of best fit.

The line slopes downwards (as t increases) so the gradient is obviously negative. Note that points (2, 25) and (3, 16) should not be used to calculate the gradient as they do not lie on the line.

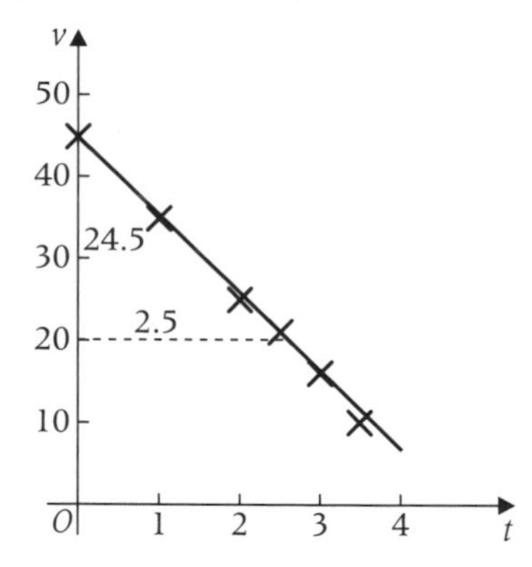

Reading exact values on a plot can be very difficult but the accuracy will slightly affect the calculated value of m. Examiners will have a range of possible values when marking; if a value lies inside the range it will be allowed full credit.

The gradient $m = \dfrac{20.5 - 45}{2.5 - 0} = \dfrac{-24.5}{2.5} = -9.8$.

The constant $c = v$-intercept $= 45$.

The equation of the line of best fit is $v = -9.8t + 45$.

(b) When $v = 0$, $\quad 0 = -9.8t + 45 \quad \Rightarrow \quad t = \dfrac{45}{9.8} = 4.59$ (to 3 s.f.).

Sometimes the scaling on the horizontal axis is such that it is not easy to read off the y-intercept. The next worked example shows you how to deal with such cases.

Worked example 7.3

Two quantities x and y are measured experimentally and the following values obtained:

x	150	160	170	180	190
y	9.0	6.9	5.0	3.1	1.0

It is thought that they are connected by a law of the form $y = ax + b$. Test if this is so and, by drawing a suitable straight line graph, estimate the values of a and b.

Solution

When you plot the points they lie in a straight line so y and x are connected by a law of the form $y = mx + c$, or $y = ax + b$, where a is the gradient of the line and b is the y-intercept (the value of y when $x = 0$).

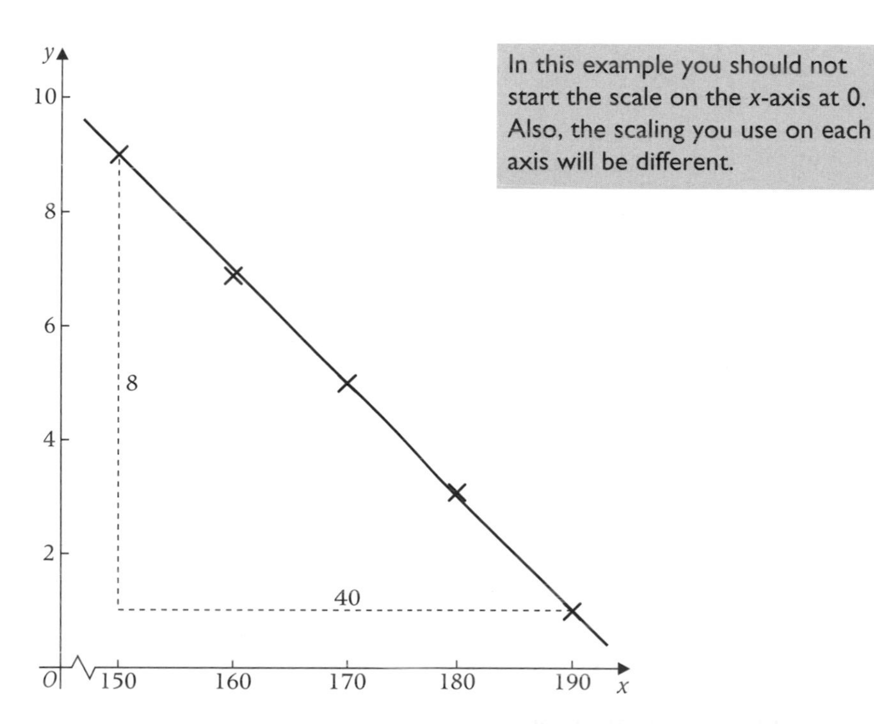

In this example you should not start the scale on the x-axis at 0. Also, the scaling you use on each axis will be different.

Using the points (150, 9) and (190, 1) which lie on the line you get

The gradient $a = \dfrac{1 - 9}{190 - 150} = \dfrac{-8}{40} = -0.2$, so $y = -0.2x + b$.

The point (170, 5) lies on the line $y = -0.2x + b$,

so $5 = -0.2 \times 170 + b \Rightarrow b = 39$.

The law connecting y and x is $y = -0.2x + 39$.

The graph does not show the value of y when $x = 0$ (since the scale on the x-axis starts at 150 not 0). In such cases you use any point on the line and substitute its coordinates into the equation of the line to find the value of b.

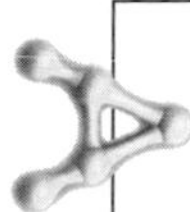

If the graph does not show the y-intercept, you can find the value of the gradient m as usual and then find the value of c by substituting the coordinates of a point on the line into the equation $y = mx + c$. Alternatively, you can use the coordinates of two points on the line to form and solve a pair of simultaneous equations in m and c.

EXERCISE 7A

1 The scatter diagram shows the results of a scientific experiment between two variables y and x. A line of best fit has been drawn.

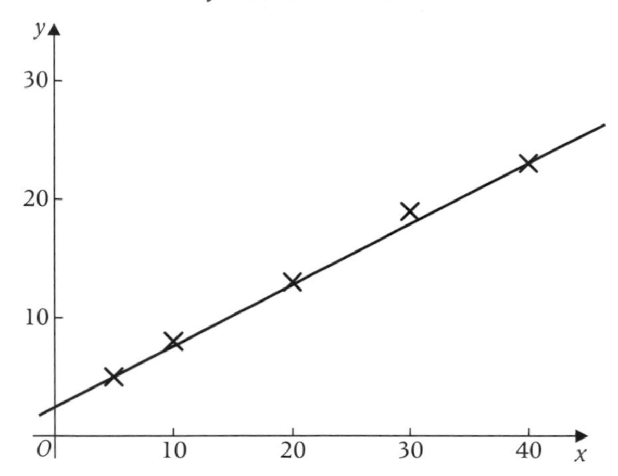

(a) Find the equation of the line of best fit.

(b) Use your equation to estimate the value of x when $y = 40$.

2 The scatter diagram shows the results of a scientific experiment between two variables S and t. A line of best fit has been drawn.

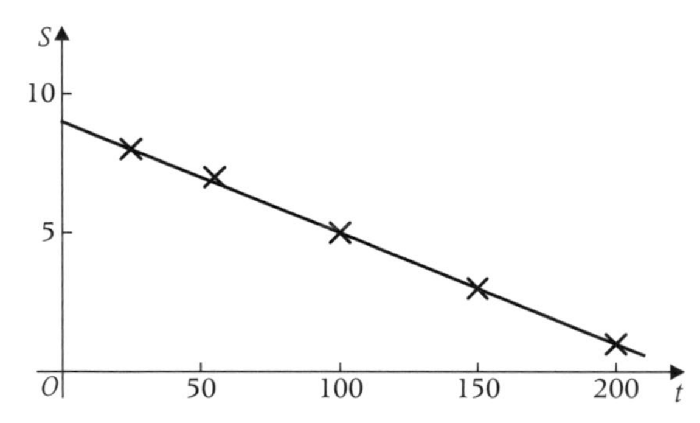

(a) Find the equation of the line of best fit.

(b) Use your equation to estimate the value of t when $S = 0$.

3 The scatter diagram shows the results of a scientific experiment between two variables R and L. A line of best fit has been drawn.

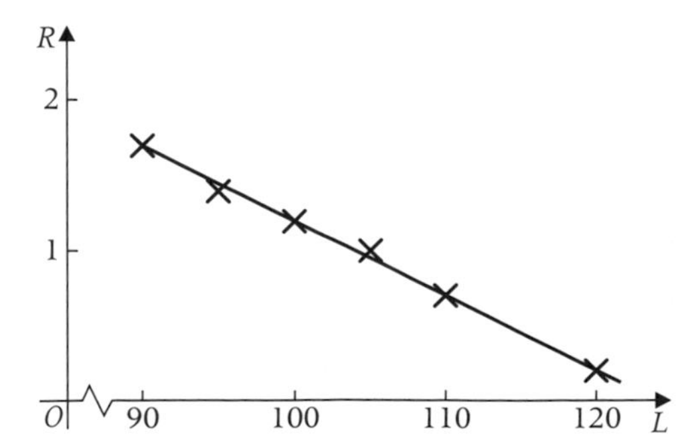

(a) Find the equation of the line of best fit.

(b) Use your equation to estimate the value of R when $L = 84$.

4 Two quantities x and y are measured experimentally. The values are shown in the table. It is thought that they are connected by a law of the form $y = px + q$. By plotting y against x show that the law is true and by drawing a suitable straight line graph, estimate the values of the constants p and q.

The phrase 'plotting y against x' implies that the vertical axis is labelled y and the horizontal axis is labelled x.

(a)

x	1	2	4	5	7
y	7.0	10.9	19.0	23.0	31.1

(b)

x	0.4	0.6	1.0	1.4	1.8
y	6.8	6.2	4.9	3.8	2.6

(c)

x	200	210	220	230	240
y	10.05	8.5	6.94	5.5	4.0

7.2 Reducing a relation to a linear law

You have seen in C1 chapter 4 that the graph of $y = x^2 + 1$, obtained by plotting y against x, is a parabola. In this section you will be shown how to write some relationships, including relations of the type $y = ax^2 + b$, in linear form. In each case, for two variables Y and X, when Y is plotted against X, a straight line graph will be obtained. As before, the general equation of the line is $Y = mX + c$, where m is the gradient and c is the Y-intercept. It is important to recognise that in general the equation $Y = mX + c$ has three terms; two of the terms (Y and mX) contain variables and the remaining term is constant (which could be zero). Throughout this section a and b will be used to represent constants.

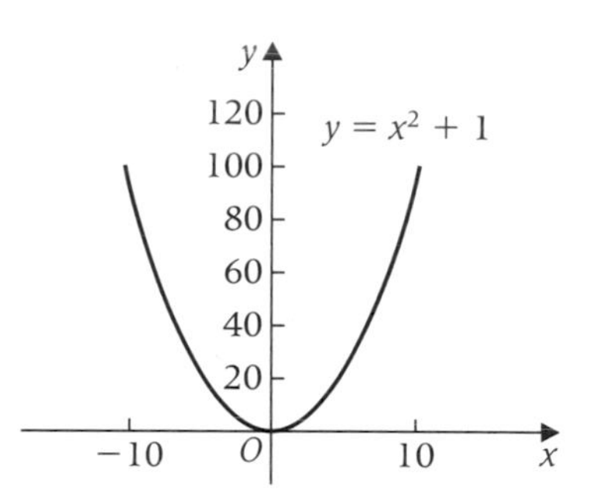

Equations of the form $y = ax^2 + b$

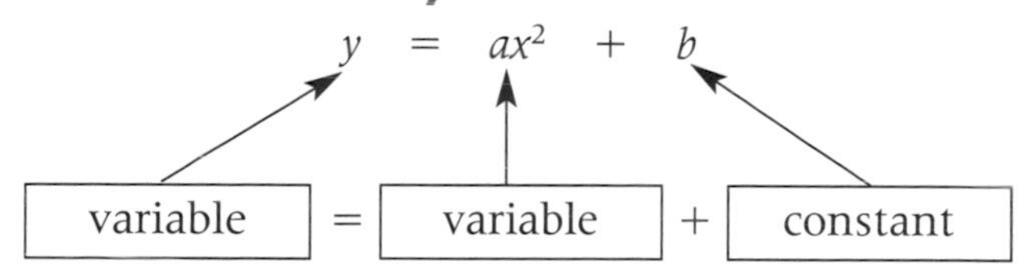

The equation $y = ax^2 + b$ is of the correct form since it has three terms, two of which contain variables and the remaining term is constant.

Comparing

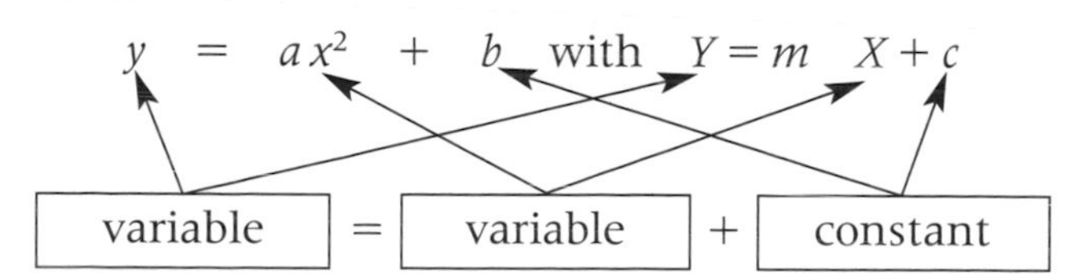

you can see that $Y = y$, $X = x^2$, $m = a$ and $c = b$.

So, to reduce the relation $y = ax^2 + b$ to a linear form you will need to plot y against x^2.

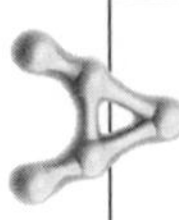

To test a belief that the relation between x and y is of the form $y = ax^2 + b$, you need to plot y against x^2. If the points are roughly in a straight line, you can deduce that the relation between x and y is of the form $y = ax^2 + b$. The gradient of the line gives an estimate for a and the intercept on the vertical axis ($X = 0$) gives an estimate for b.

Worked example 7.4

The table shows some experimental values of the variables L and T.

T	5	10	12	15	20
L	11	29	41	61	105

A scientist believes that the variables T and L satisfy a relation of the form $L = aT^2 + b$.

(a) By drawing an appropriate graph, explain why the scientist is correct.

(b) Use your graph to estimate the values of the constants a and b.

(c) Given that $T > 0$, use the relation to find the value of T when $L = 149$.

Solution

(a) Comparing

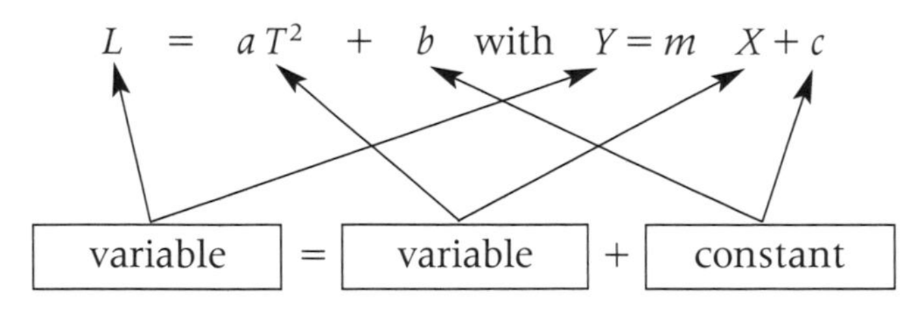

you can see that $Y = L$, $X = T^2$, $m = a$ and $c = b$,

so plotting L against T^2 should lead to a straight line graph if the relationship $L = aT^2 + b$ is correct.

The table of values for L and T^2 is

T^2	25	100	144	225	400
L	11	29	41	61	105

The graph of L against T^2 is

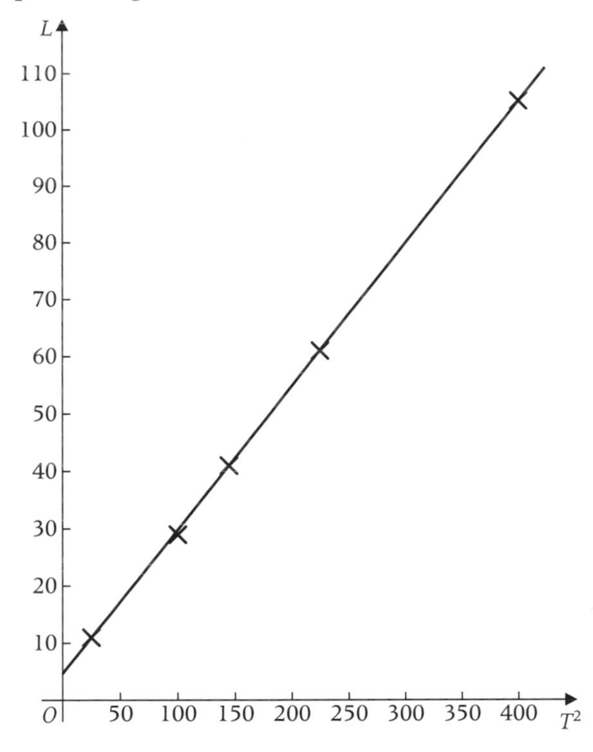

Since the points lie in a line the scientist is correct.

(b) $a = \text{gradient of line} = \dfrac{61 - 11}{225 - 25} = \dfrac{50}{200} = \dfrac{1}{4}.$

$b = \text{intercept on } L\text{-axis} = 5.$

(c) The relation is $L = \dfrac{1}{4}T^2 + 5.$

When $L = 149$, $144 = \dfrac{1}{4}T^2 \Rightarrow T = 24$ (since $T > 0$).

Equations of the form $\dfrac{1}{x} + \dfrac{1}{y} = a$

Rearranging the equation $\dfrac{1}{x} + \dfrac{1}{y} = a$ you get

$$\frac{1}{y} = -\frac{1}{x} + a$$

variable = variable + constant

The equation $\dfrac{1}{y} = -\dfrac{1}{x} + a$ is of the correct form since it has three terms, two of which contain variables and the remaining term is constant.

Comparing

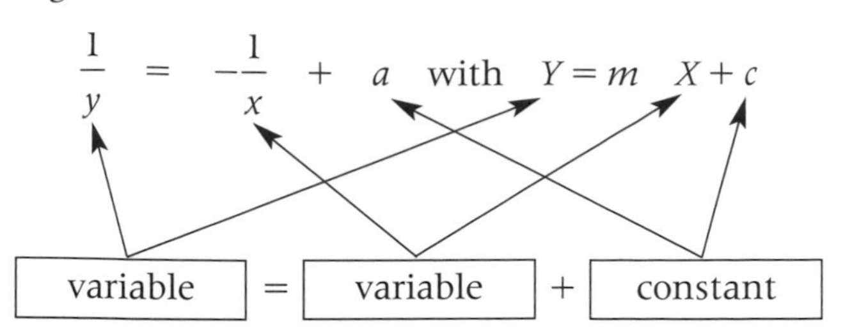

you can see that $Y = \frac{1}{y}$, $X = \frac{1}{x}$, $m = -1$ and $c = a$,

so, to reduce the relation $\frac{1}{x} + \frac{1}{y} = a$ to a linear form you will need to plot $\frac{1}{y}$ against $\frac{1}{x}$.

To test a belief that the relation between x and y is of the form $\frac{1}{x} + \frac{1}{y} = a$, you need to plot $\frac{1}{y}$ against $\frac{1}{x}$. If the points are roughly in a straight line with gradient -1, you can deduce that the relation between x and y is of the form $\frac{1}{x} + \frac{1}{y} = a$. The intercept on the vertical axis ($X = 0$) gives an estimate for a.

Worked example 7.5

The table shows some experimental values of the variables x and y.

x	10	15	25	40	50
y	11.1	8.1	6.7	6.1	5.9

(a) By plotting $\frac{1}{y}$ against $\frac{1}{x}$ show that these values are consistent with the relation $\frac{1}{x} + \frac{1}{y} = a$, where a is a constant.

(b) Estimate the value of a, giving your answer to two decimal places.

(c) Hence estimate the value of y when $x = 0.5$, giving your answer to two decimal places.

Solution

(a) Rearranging the equation $\frac{1}{x} + \frac{1}{y} = a$ you get

$$\frac{1}{y} = -\frac{1}{x} + a.$$

Plotting $\frac{1}{y}$ against $\frac{1}{x}$ should lead to a straight line graph with gradient -1. The intercept on the $\frac{1}{y}$-axis gives the value of a. The values, rounded to three decimal places, to be plotted are:

$\frac{1}{x}$	0.1	0.067	0.04	0.025	0.02
$\frac{1}{y}$	0.090	0.123	0.149	0.164	0.169

The graph is

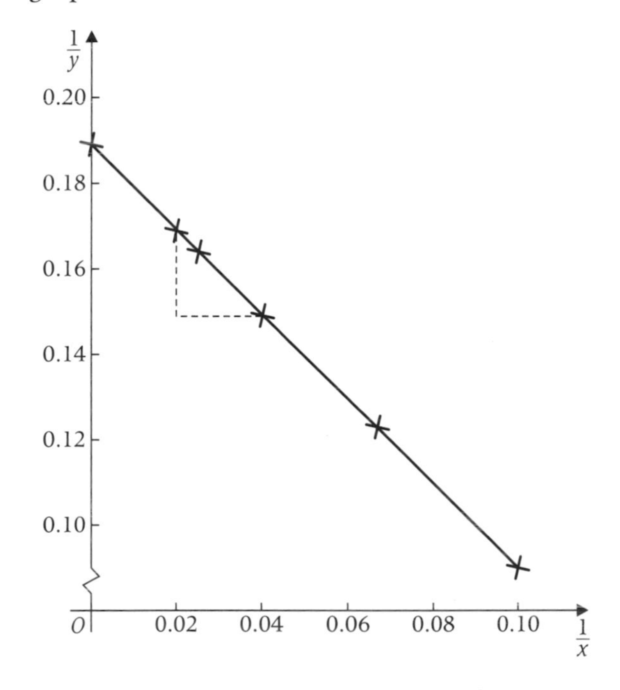

The gradient of the line is $\frac{0.149 - 0.169}{0.04 - 0.02} = -1$.

You could have used the coordinates of any two points on the straight line.

Since the points lie roughly on the line with gradient -1 you can deduce that the values are consistent with the relation $\frac{1}{x} + \frac{1}{y} = a$.

(b) $a =$ intercept on the $\frac{1}{y}$-axis so $a = 0.19$.

(c) When $x = 0.5$, $\frac{1}{x} = 2$.

Using $\frac{1}{x} + \frac{1}{y} = a$, you get $2 + \frac{1}{y} = 0.19$

$\Rightarrow \quad \frac{1}{y} = -1.81 \quad \Rightarrow \quad y = \frac{1}{-1.81} = -0.55$ (to 2 d.p.).

Use the relation $\frac{1}{x} + \frac{1}{y} = a$, since the scaling on the $\frac{1}{x}$-axis does not go as far as 2.

Equations of the form $y = ax^2 + bx$

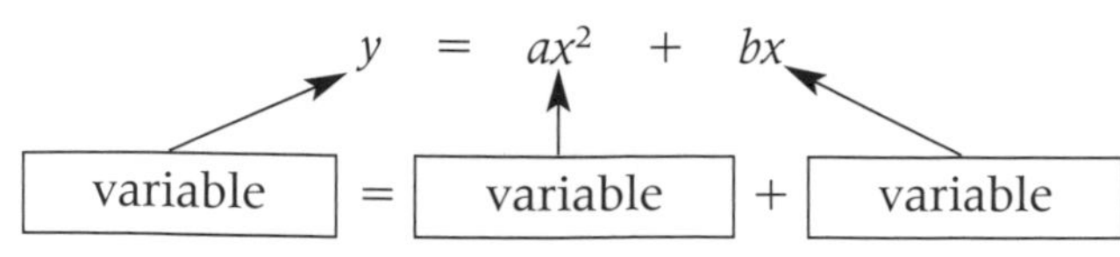

For $y = ax^2 + bx$, if you plot y against x, the graph is a parabola through the origin.

The equation $y = ax^2 + bx$ is **not** of the correct form since it has three variable terms and no constant term.

To overcome this problem you must rearrange the equation so that it is of a linear form, that is an equation with two variables and one constant term. There are many ways in which this can be done.

You could have divided each term by x^2 to get a linear form, or divided each term by y.

For example, you can divide each term by x to get $\frac{y}{x} = ax + b$.

Comparing

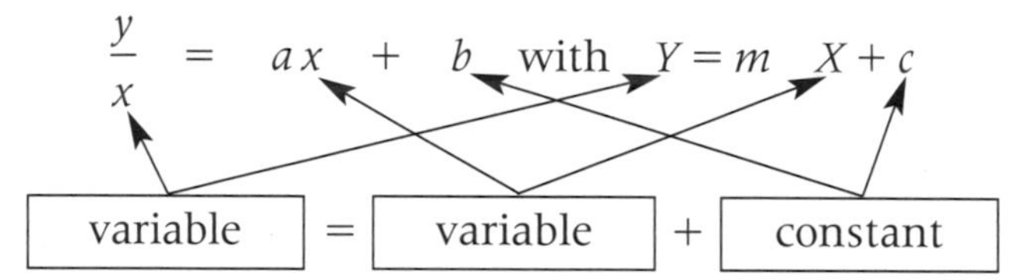

you can see that $Y = \frac{y}{x}$, $X = x$, $m = a$ and $c = b$,
so, to reduce the relation $y = ax^2 + bx$ to a linear form you could plot $\frac{y}{x}$ against x.

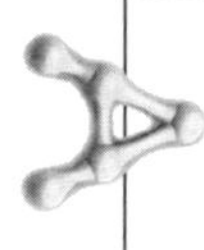

To test a belief that the relation between x and y is of the form $y = ax^2 + bx$, you can plot $\frac{y}{x}$ against x. If the points are roughly in a straight line, you can deduce that the relation between x and y is of the form $y = ax^2 + bx$. The gradient of the line gives an estimate for a and the intercept on the vertical axis ($X = 0$) gives an estimate for b.

Worked example 7.6

The table shows some experimental values of the variables x and the corresponding values of y.

x	2	4	5	6	8
y	14	32	42	54	80

(a) By plotting $\frac{y}{x}$ against x show that these values are consistent with the relation $y = ax^2 + bx$, where a and b are constants.

(b) Draw a suitable straight line to illustrate the relation and use your line to estimate the value of y when $x = 7$.

(c) Estimate the values of a and b, giving your answers to one decimal place.

Solution

(a) Calculating $\frac{y}{x}$ the table of values becomes:

x	2	4	5	6	8
$\frac{y}{x}$	7	8	8.4	9	10

When these values are plotted the graph shows that the points all approximately lie in a straight line. The equation of the line is of the form $\frac{y}{x} = ax + b$, where a is the gradient of the line and b is the intercept on the $\frac{y}{x}$-axis.

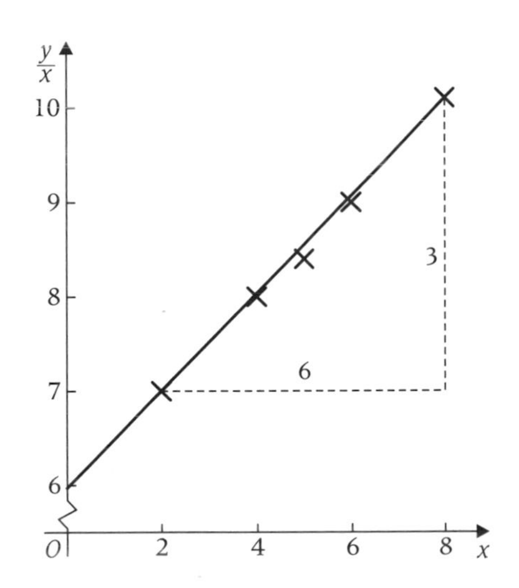

Multiplying each term of the equation $\frac{y}{x} = ax + b$ by x you get the relation $y = ax^2 + bx$.

(b) From the graph, when $x = 7$, $\frac{y}{x} = 9.5 \Rightarrow y = 7 \times 9.5 = 66.5$.

(c) $a =$ gradient of line $= \dfrac{10 - 7}{8 - 2} = 0.5$

$b =$ intercept on $\frac{y}{x}$-axis (i.e. the line $x = 0$) $= 6$.

Worked example 7.7

For each of the following relations between the variables x and y, find possible variables which can be plotted to obtain a straight line graph and explain how the graph can be used to estimate the value of the constant a and the value of the constant b.

(a) $ay^2 = x - b$ **(b)** $y^3 = ax^2 + bx$

Solution

(a) $ay^2 = x - b$ has three terms, two contain variables and the remaining term is constant. This is a linear form. Rearranging the relation so that it matches $Y = mX + c$ gives $x = ay^2 + b$.

Comparing gives $Y = x$, $X = y^2$, $m = a$ and $c = b$.

Plotting x against y^2 should give a straight line graph.

a is the gradient of the line and b is the intercept on the x-axis.

Note that the vertical axis will be labelled x and the horizontal axis will be labelled y^2.

(b) $y^3 = ax^2 + bx$ has three variable terms so it is **not** a linear form. You must rewrite the relationship so that there are two variable terms and one constant term. Dividing each term by x is one possible way. This gives $\frac{y^3}{x} = ax + b$ which compares with $Y = mX + c$ to give $Y = \frac{y^3}{x}$, $X = x$, $m = a$ and $c = b$.

Plotting $\frac{y^3}{x}$ against x should give a straight line graph.

a is the gradient of the line and b is the intercept on the $\frac{y^3}{x}$-axis.

Dividing each term by x^2 or dividing each term by ay^3 are two other ways to obtain a linear form.

EXERCISE 7B

1 For each of the following relations between the variables x and y, find possible variables which can be plotted to obtain a straight line graph and explain how the graph can be used to estimate the value of the constant a and the value of the constant b:

(a) $y = ax^3 + b$ **(b)** $y = a + b\sqrt{x}$ **(c)** $y^2 = ax + b$

(d) $\frac{1}{y} = a + \frac{b}{\sqrt{x}}$ **(e)** $y = ax^3 + bx$ **(f)** $y = ax + by^2$

(g) $y = \frac{a}{x} + bx$ **(h)** $xy = ax^2 + b$ **(i)** $y^2 - ay + bx^2 = 0$

2 The diagram shows a straight line graph of $\frac{y}{x}$ against x passing through the points (0, 2) and (2, 5).

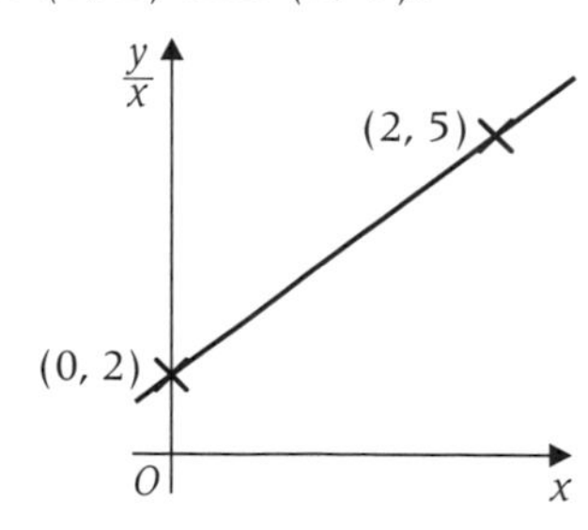

(a) Express y in the form $ax^2 + bx$, where a and b are constants to be found.

(b) Hence verify that $y = 47.5$ when $x = 5$.

3 It is assumed that x and y are related by a law of the form $y = a + bx^2$, where a and b are constants. Experimental measurements of x and y are taken to give the following pairs of values:

x	8	10	12	14	16
y	40	60	82	108	138

(a) By means of a straight line graph verify that the law is valid.

(b) Use your graph to estimate approximate values for a and b.

4 The table shows corresponding values of the variables x and y obtained in an experiment.

x	2.5	3	3.5	4	4.5
y	1.020	0.955	0.914	0.885	0.864

(a) Draw a straight line graph to verify that x and y are approximately connected by a relation of the form $\frac{1}{x} + \frac{1}{y} = a$, where a is a constant.

(b) Use your graph to estimate the value, to two decimal places, of

(i) x when $y = 0.9$, **(ii)** a.

5 The diagram shows a straight line graph of xy against x passing through the points (1, 3) and (5, 1).

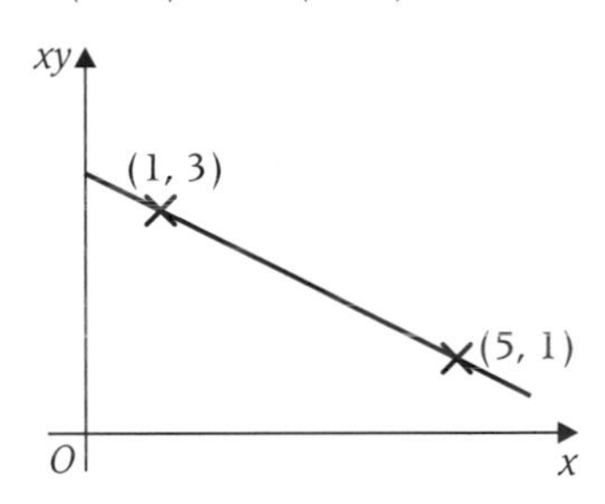

(a) Express y in the form $\frac{a}{x} + b$, where a and b are constants to be found.

(b) Hence verify that $y = -0.43$ when $x = 50$.

6 The variables x and y are known to satisfy an equation of the form $y = a + b\sqrt{x}$, where a and b are constants. Corresponding approximate values of x and y (each rounded to one decimal place) were obtained experimentally and are given in the following table.

x	3.2	6.8	16.0	25.2	33.6	40.4
y	4.0	5.0	6.6	8.2	9.0	9.8

By drawing a suitable linear graph, estimate the values of a and b, giving both answers to one decimal place. [A]

7.3 The use of logarithms to reduce equations of the form $y = ax^n$ to a linear law

In many cases scientists know that two variables are directly proportional to each other, when one variable is equal to zero the other variable is zero. They start with the assumption that the relation between the variables x and y is likely to be of the form $y = ax^n$, where a and n are constants, and then apply experimental data to find the value of the constants. In this section you will be shown how to write the relation $y = ax^n$ in a linear form by using the laws of logarithms and then you will be shown how to find estimates for the constants a and n.

See C2 section 11.3 for the laws of logarithms.

Consider the relationship $y = ax^n$ between the variables x and y. To reduce this relationship to linear form you:

Since an unknown occurs as a power, you take logarithms.

1 take logarithms of both sides $\Rightarrow$ $\log y = \log ax^n$

2 apply the first law of logarithms $\Rightarrow$ $\log y = \log a + \log x^n$

3 apply the third law of logarithms $\Rightarrow$ $\log y = \log a + n \log x$

Note that $ax^n = a \times x^n$, (n is not the power of a).

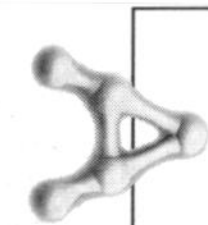

Taking logarithms of both sides of
$y = ax^n \Rightarrow \log y = \log a + n \log x$.

The equation $\log y = \log a + n \log x$ is of the correct linear form since it has three terms, two of which contain variables and the remaining term is constant.

Comparing

$\log y = n \log x + \log a$ with $y = m \;\; X + c$

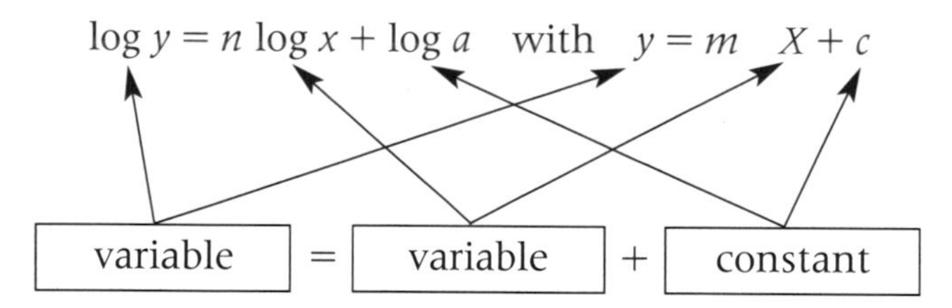

you can see that $Y = \log y$, $X = \log x$, $m = n$ and $c = \log a$, so to represent the relation $y = ax^n$ in a linear form you need to plot $\log y$ against $\log x$.

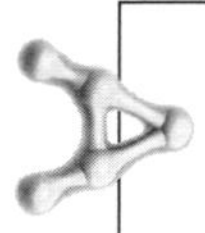

To test a belief that the relation between x and y is of the form $y = ax^n$, you need to plot $\log y$ against $\log x$. If the points are roughly in a straight line, you can deduce that the relation between x and y is of the form $y = ax^n$. The gradient of the line gives an estimate for n and the intercept on the vertical axis ($X = 0$) gives the value of $\log a$ from which the estimate for a can be found.

Worked example 7.8

The diagram shows a straight line graph of $\log_{10}y$ against $\log_{10}x$ passing through the points (0, 3) and (4, 6).

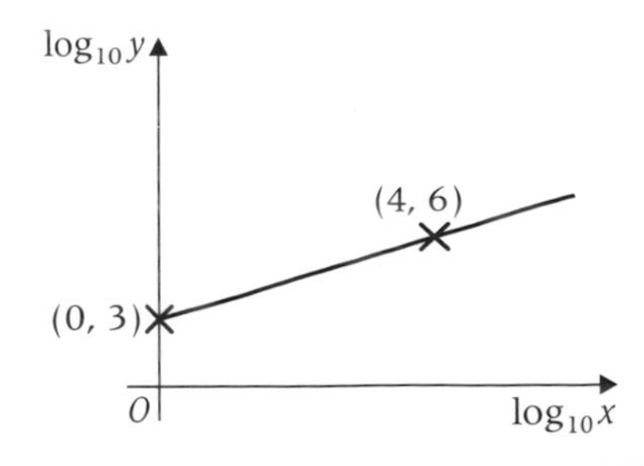

Express y in the form ax^b, where a and b are constants to be found.

It is important to realise that (4, 6) refers to ($\log_{10}x$, $\log_{10}y$) not (x, y).

Solution

The equation of the line takes the form $Y = mX + c$, where $Y = \log_{10}y$, $X = \log_{10}x$, m is the gradient and c is the intercept on the Y-axis.

$$m = \frac{6-3}{4-0} = 0.75 \quad \text{and} \quad c = 3.$$

The equation of the line is $\log_{10}y = 0.75\log_{10}x + 3$

$$\Rightarrow \log_{10}y = 0.75\log_{10}x + 3\log_{10}10$$

$$\Rightarrow \log_{10}y = \log_{10}x^{0.75} + \log_{10}10^3$$

$$\Rightarrow \log_{10}y = \log_{10}(x^{0.75} \times 10^3)$$

$$\Rightarrow \log_{10}y = \log_{10}(1000x^{0.75})$$

$$\Rightarrow y = 1000x^{0.75}.$$

So $y = ax^b$, where $a = 1000$ and $b = 0.75$.

7

Worked example 7.9

The corresponding values of two variables x and t found by experiment are:

t	2	4	6	8	10
x	3.16	17.9	49.3	101.2	176.8

By drawing a suitable linear graph, verify that the values of t and x, approximately satisfy a relation of the form $x = at^n$. Use your graph to estimate values of the constants a and n giving your answers to two significant figures.

Solution

Since n is unknown and occurs as a power you need to take logarithms of both sides

$$\Rightarrow \log x = \log at^n$$

$$\Rightarrow \log x = \log a + \log t^n$$

$$\Rightarrow \log x = \log a + n\log t$$

$\log x = n\log t + \log a$ is of the form $Y = mX + c$, so plotting $\log x$ against $\log t$ should give a straight line graph. The gradient of the line gives n and the intercept on the $\log x$-axis gives $\log a$.

By using logarithms to base 10, the table of values, to three s.f. becomes:

$\log_{10} t$	0.301	0.602	0.778	0.903	1.00
$\log_{10} x$	0.500	1.25	1.69	2.01	2.25

You could use natural logarithms (the ln button on a calculator) instead of base 10. The values for a and n should work out the same.

When these values are plotted the graph shows that the points all approximately lie in a straight line which verifies the relationship $x = at^n$.

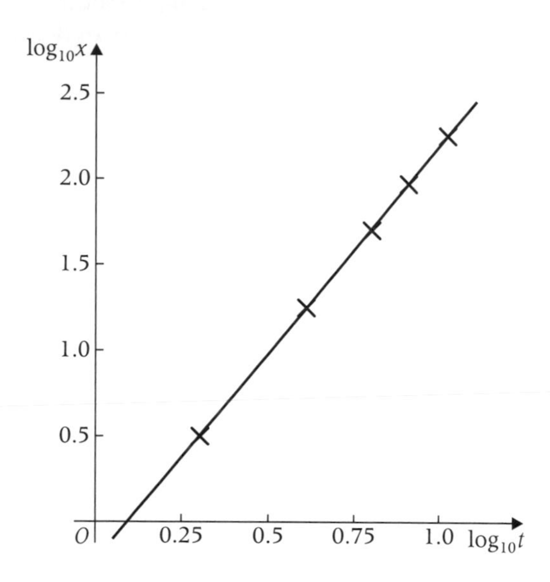

$$n = \text{gradient of line} = \frac{1.25 - 0.50}{0.602 - 0.301} = \frac{0.75}{0.301} = 2.5 \text{ (2 s.f.)}$$

Points (0.602, 1.25) and (0.301, 0.5) lie on the line.

$$\log_{10} a = -0.25 \quad \Rightarrow \quad a = 10^{-0.25} = 0.56 \text{ (2 s.f.)}$$

EXERCISE 7C

1 The diagram shows a straight line graph of $\log_{10} y$ against $\log_{10} x$ passing through the points (0, 0.5) and (4, 3.5).

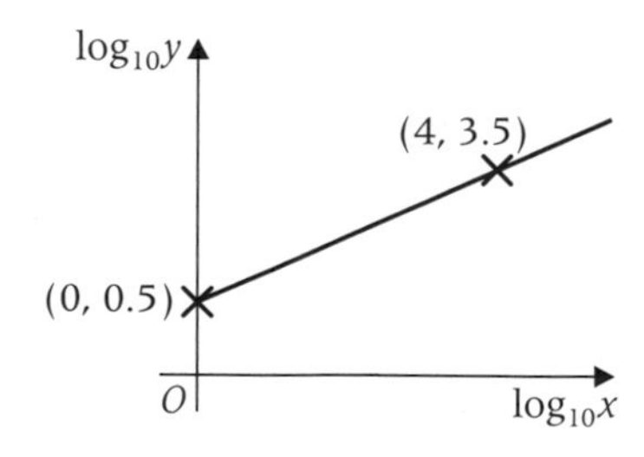

Express y in the form ax^b, where a and b are constants to be found.

2 The diagram shows a straight line graph of $\log_{10}y$ against $\log_{10}x$ passing through the points (2, 2) and (5, 8).

Express y in the form ax^b, where a and b are constants to be found.

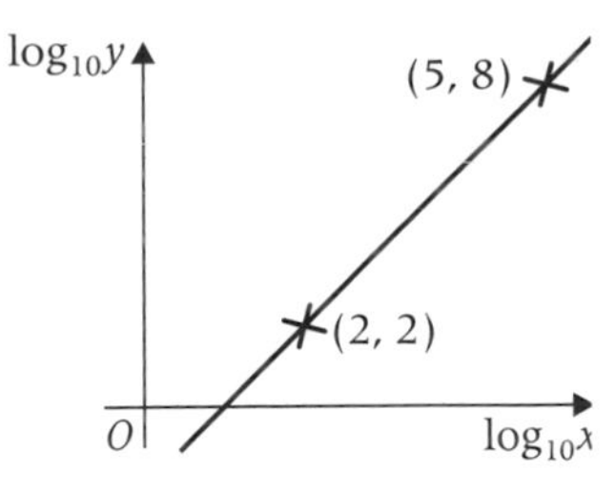

3 The corresponding pairs of values of two variables y and x found by experiment are:

x	2	3	4	5	6
y	9.2	14.9	21.1	27.6	34.4

By drawing a suitable linear graph, verify that the values of x and y, approximately satisfy a relation of the form $y = ax^b$. Use your graph to estimate values of the constants a and b giving your answers to two significant figures.

4 V is thought to relate T by a law of the form $V = aT^{-n}$, where a and n are constants.

(a) Express $\log_{10}V$ in terms of n, $\log_{10}T$ and $\log_{10}a$.

Pairs of values of T and V are:

T	10	12	14	16	18
V	0.980	0.895	0.829	0.775	0.731

(b) Plot $\log_{10}V$ against $\log_{10}T$ and hence draw a suitable straight line to illustrate the relation between the data.

(c) Use your line to estimate, to two significant figures:
(i) the value of V when $T = 12.6$,
(ii) the values of a and n.

5 The variables Q and x satisfy a relation of the form $Q = ax^b$, where a and b are constants.

Measurements of Q for given values of x gave the following results:

x	0.4	0.5	0.6	0.7	0.8
Q	1.72	3.02	4.74	6.98	9.73

(a) Express $\log_{10}Q$ in terms of $\log_{10}a$, b and $\log_{10}x$.

(b) (i) Plot $\log_{10}Q$ against $\log_{10}x$.
(ii) Draw a suitable straight line to illustrate the relation between the data.

(c) Use your line to estimate:
(i) the value of Q when $x = 0.54$, giving your answer to two significant figures,
(ii) the values of a and b, giving your answer to two significant figures. [A adapted]

7.4 The use of logarithms to reduce equations of the form $y = ab^x$ to a linear law

The equation $y = ab^x$, where a and b are constants, has a variable power. In this section, once again you will use logarithms to reduce the relation to linear form.

Consider the relation $y = ab^x$ between the variables x and y. To reduce this relation into linear form you:

Note that $ab^x = a \times b^x$, (x is not the power of a).

1 take logarithms of both sides $\Rightarrow$ $\log y = \log ab^x$
2 apply the first law of logarithms $\Rightarrow$ $\log y = \log a + \log b^x$
3 apply the third law of logarithms $\Rightarrow$ $\log y = \log a + x \log b$

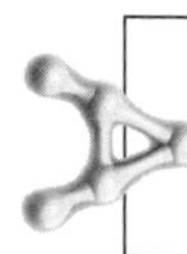

Taking logarithms of both sides of
$y = ab^x \quad \Rightarrow \quad \log y = \log a + x \log b$.

The equation $\log y = \log a + x \log b$ is of the correct linear form since it has three terms, two of which contain variables and the remaining term is constant.

Comparing

$$\log y = (\log b)x + \log a \quad \text{with} \quad Y = mX + c$$

variable = variable + constant

you can see that $Y = \log y$, $X = x$, $m = \log b$ and $c = \log a$.

So,

To represent the relation $y = ab^x$ in a linear form you need to plot $\log y$ against x. If a straight line is obtained from the given data the relation is true. The gradient of the line is the value of $\log b$ and the intercept on the vertical axis ($X = 0$) gives the value of $\log a$. Knowing these two values, estimates for a and b can be found.

Worked example 7.10

The diagram shows a straight line graph of $\log_{10} y$ against x passing through the points (0, 4) and (5, 1).

Express y in the form ab^x, where a and b are constants to be found.

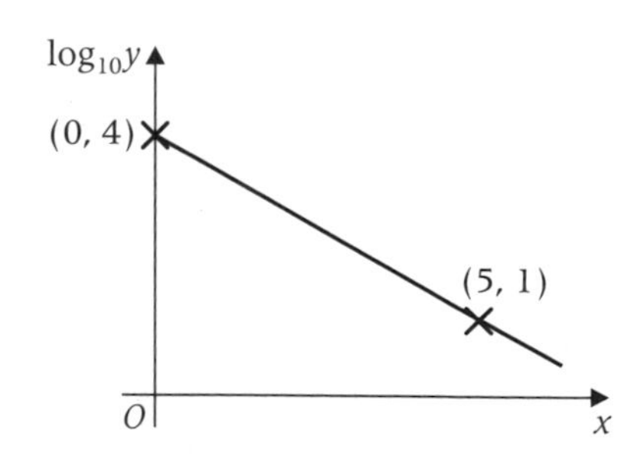

Solution

The equation of the line takes the form $Y = mX + c$, where $Y = \log_{10} y$, $X = x$, m is the gradient and c is the intercept on the Y-axis.

$$m = \frac{1 - 4}{5 - 0} = -0.6 \quad \text{and} \quad c = 4$$

The equation of the line is $\log_{10}y = -0.6x + 4$

$$\Rightarrow \quad y = 10^{-0.6x + 4}$$
$$\Rightarrow \quad y = 10^{-0.6x} \times 10^4$$
$$\Rightarrow \quad y = 10\,000 \times (10^{-0.6})^x$$
$$\Rightarrow \quad y = 10\,000 \times (0.251\ldots)^x$$

So $y = ax^b$, where $a = 10\,000$ and $b = 0.251$ (3 s.f.).

Worked example 7.11

The data for x and y as given in the table below are related approximately by a law of the form $ky = h^x$, where h and k are constants.

x	1	2	3	4	5	6
y	17	49	110	330	810	2200

By drawing a suitable graph find estimates, to two significant figures, for h and k. [A]

Solution

To reduce the relation $ky = h^x$ to linear form you need to take logarithms to base 10 of both sides

Alternatively, you could have taken ln of both sides.

$$\Rightarrow \quad \log_{10}(ky) = \log_{10}(h^x)$$
$$\Rightarrow \quad \log_{10} k + \log_{10} y = x \log_{10} h$$

Used the first and the third laws of logarithms.

Re-writing this as $\log_{10} y = (\log_{10} h)x - \log_{10} k$ and comparing with $Y = mX + c$ you can see that it is of linear form with $Y = \log_{10}y$, $X = x$, $m = \log_{10}h$ and $c = -\log_{10}k$. You need to draw a graph of $\log_{10}y$ against x.

Produce a relevant table of values. Since answers are expected to 2 s.f. you must show values for $\log_{10}y$ to at least 3 s.f. in the table.

x	1	2	3	4	5	6
$\log_{10}y$	1.23	1.69	2.04	2.52	2.91	3.34

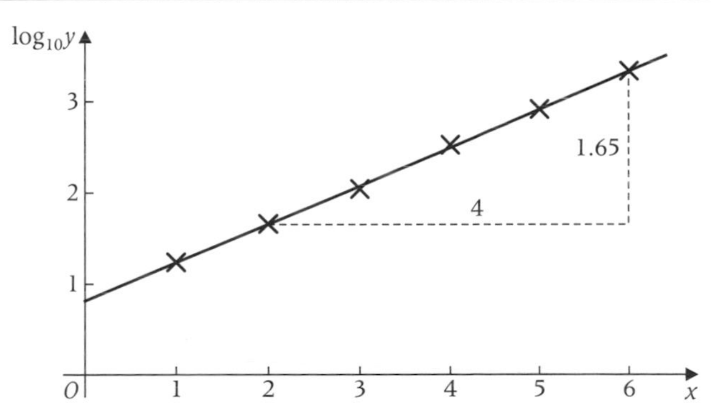

Always use sensible scaling on the axes. This should make plotting points a simple process.

$$\text{Gradient} = \frac{3.34 - 1.69}{6 - 2} = \frac{1.65}{4} = 0.4125$$

$$\log_{10}h = 0.4125 \quad \Rightarrow \quad h = 10^{0.4125} = 2.58\ldots$$

Intercept on the Y-axis $= 0.84$

$$-\log_{10}k = 0.84 \Rightarrow \log_{10}k = -0.84 \Rightarrow k = 10^{-0.84} = 0.1445\ldots$$

So, to two significant figures, $h = 2.6$ and $k = 0.14$.

Acceptable 2 s.f. answers in the exam:
h: 2.5 to 2.8
k: 1.2 to 1.6

Worked example 7.12

This worked example shows you how unknown constants can be found by forming and solving a pair of simultaneous equations.

Given that $y = ab^x$, where a and b are constants, express $\log y$ in terms of $\log a$, $\log b$ and x. Hence, given that $y = 700$ when $x = 3$ and $y = 1100$ when $x = 5$, find, to three significant figures, the value of a and the value of b.

Solution

$y = ab^x \quad \Rightarrow \quad \log y = \log(ab^x) = \log a + \log b^x = \log a + x \log b$

When $x = 3$, $y = 700 \quad \Rightarrow \quad \log_{10} 700 = \log_{10} a + 3 \log_{10} b \quad [1]$

When $x = 5$, $y = 1100 \quad \Rightarrow \quad \log_{10} 1100 = \log_{10} a + 5 \log_{10} b \quad [2]$

$[2] - [1] \quad \Rightarrow \quad \log_{10} 1100 - \log_{10} 700 = 2 \log_{10} b$

$$\Rightarrow \quad \log_{10} b = \frac{1}{2}\log_{10} \frac{1100}{700} = 0.098\,147\ldots$$

$$\Rightarrow \quad b = 10^{0.098\,14\ldots} \quad \Rightarrow \quad b = 1.253\,56\ldots = 1.25 \text{ (3 s.f.)}$$

$5 \times [1] - 3 \times [2]$ gives $5 \log_{10} 700 - 3 \log_{10} 1100 = 2 \log_{10} a$

$$\Rightarrow \quad \log_{10} a = \frac{1}{2}\log_{10} \frac{700^5}{1100^3} = 2.550\,656\ldots$$

$$\Rightarrow \quad a = 10^{2.550\,656\ldots} \quad \Rightarrow \quad a = 355.3497\ldots = 355 \text{ (3 s.f.)}$$

To compare the two methods you should try answering this question by plotting $\log_{10} y$ against x for the two given points. In general, students' solutions are less prone to error in the graphical approach.

EXERCISE 7D

1 The diagram shows a straight line graph of $\log_{10} y$ against x passing through the points (0, 0.5) and (5, 3).

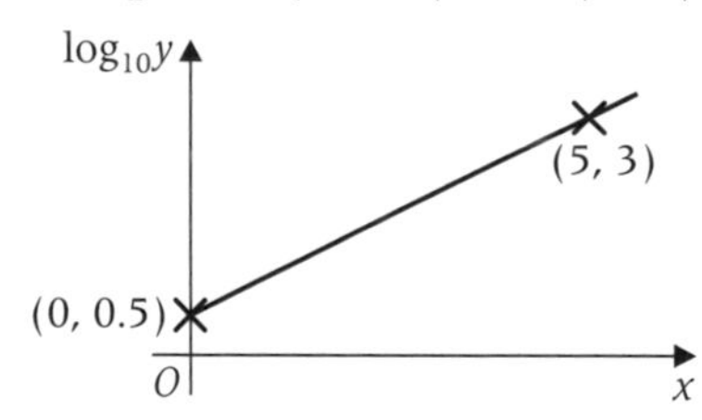

Express y in the form ab^x, where a and b are constants to be found.

2 The diagram shows a straight line graph of $\log_{10} P$ against T passing through the points (0, 1) and (5, 0.25).

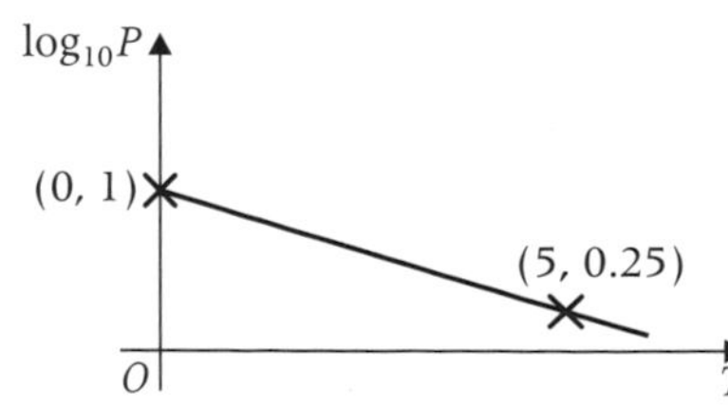

Express P in the form ab^T, where a and b are constants to be found.

3 T is thought to relate L by a law of the form $T = ka^L$, where k and a are constants.

(a) Express $\log_{10}T$ in terms of L, $\log_{10}k$ and $\log_{10}a$.

The pairs of values of L and T are:

L	1	2	3	4	5
T	13.95	43.24	134.06	415.58	1288.31

(b) Plot $\log_{10}T$ against L on graph paper and hence draw a suitable straight line to illustrate the relation between the data.

(c) Use your line to estimate, to two significant figures:

(i) the value of T when $L = 2.5$,

(ii) the values of k and a.

4 The data for x and y, as given in the table below, are related approximately by a law of the form $y = ab^x$, where a and b are constants.

x	0.5	1	1.5	2	2.5
y	10.6	15.0	21.2	30.0	42.4

By drawing a suitable graph find estimates, to two significant figures, for a and b.

5 The data for x and y, as given in the table below, are related approximately by a law of the form $ky = h^x$, where h and k are constants.

x	1	2	3	4	5	6
y	0.96	2.30	5.53	13.27	31.85	76.44

By drawing a suitable graph find estimates, to two significant figures, for h and k.

6 The data for x and y, as given in the table below, are related approximately by a law of the form $y = pq^{-x}$, where p and q are constants.

x	1	2	3	4	5
y	15.0	9.38	5.86	3.66	2.29

By drawing a suitable graph find estimates, to two significant figures, for p and q.

MIXED EXERCISE

Some of the following past examination questions have been adapted to stay within the MFP1 specification.

1 The variables x and y satisfy a relation of the form $y = ax^b$, where a and b are constants.

Measurements of y for given values of x gave the following results:

x	2	3	4	5	6
y	6.32	7.24	7.98	8.60	9.12

(a) Plot $\log_{10}y$ against $\log_{10}x$ and draw the line of best fit to the plotted points.

(b) Use your line to estimate:

(i) the value of x when $y = 7.50$, giving your answer to two significant figures,

(ii) the values of a and b, giving your answer to an appropriate degree of accuracy [A]

2 The variables x and y are believed to satisfy an equation of the form $\frac{1}{y} = \frac{1}{x} + \frac{1}{a}$, where a is a constant. For four chosen values of x, the corresponding approximate values of y, rounded to three significant figures, are obtained experimentally. The results are given in the following table:

x	2.5	5	7.5	10
y	4.35	28.5	−33.0	−14.9

By drawing a suitable linear graph, obtain an estimate for the value of a. [A]

3 It is assumed that x and y are related by a law of the form $y = ax^3 + bx^2$, where a and b are constants. Pairs of values of x and y are given in the table:

x	1.25	2.53	3.42	4.87	5.69
y	10.9	76.7	183.4	507.4	802.9

(a) Rearrange the equation $y = ax^3 + bx^2$ in the form $z = ax + b$, where z is an expression containing x and y.

(b) By means of a straight line graph verify that the law is valid.

(c) Use your graph to estimate approximate values for a and b.

4 The variables x and y are known to satisfy an equation of the form $y = ab^x$, where a and b are constants. For five different values of x, corresponding approximate values of y were obtained experimentally. The results are given in the following table:

x	2.0	2.5	3.0	3.5	4.0
y	11.3	18.0	27.1	44.5	70.4

By drawing a suitable linear graph, estimate the values of a and b, giving both answers to one decimal place. [A]

5 A mathematical model is required to estimate the number, N, of a certain strain of bacteria in a test tube at time t hours after a certain instant.

After values of $\log_{10}N$ are plotted against t, a straight line graph can be drawn through the points as shown below.

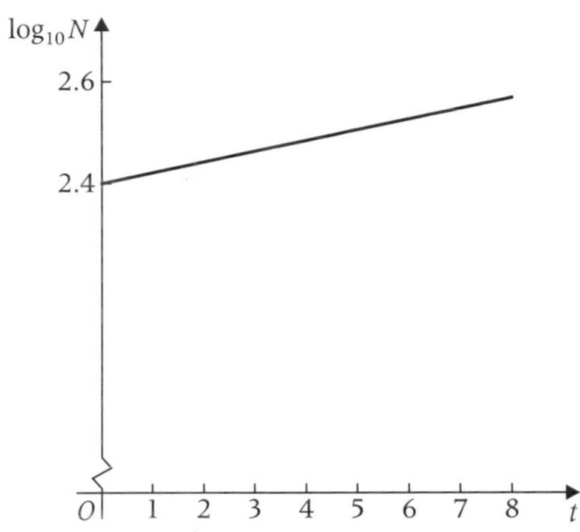

(a) Use the graph to estimate the number of bacteria when $t = 5$.

(b) The graph would suggest that N and t are related by an equation of the form $N = a \times b^t$, where a and b are constants.

(i) Express $\log_{10}N$ in terms of $\log_{10}a$, $\log_{10}b$ and t.

(ii) Use the graph to determine the values of a and b, giving your answers to three significant figures.

(c) Suggest why the model $N = a \times b^t$ is likely to give an overestimate of the number of bacteria in the test tube for large values of t. [A]

6 The following measurements of the volume, $V\,\text{cm}^3$, and the pressure, p cm of mercury, of a given mass of gas were taken.

V	10	50	110	170	230
p	1412.5	151.4	50.3	27.4	18.6

(a) By plotting $\log_{10}p$ against $\log_{10}V$, verify graphically the relation $p = kV^n$, where k and n are constants.

(b) Use your graph to find approximate values for k and n, giving your answer to two significant figures. [A]

Key point summary

1 Since experimental data is not exact, due to measuring errors, points plotted are unlikely to all lie exactly in line so a line of best fit is drawn. *p86*

2 $y = mx + c$ is the equation of a straight line with gradient m and y-intercept c. *p86*

7

3 If the variables used are not x and y, the method to find the equation of the line of best fit is exactly the same. In the general equation $y = mx + c$ you just replace y by the variable on the vertical axis and replace x by the variable on the horizontal axis. p87

4 If the graph does not show the y-intercept, you can find the value of the gradient m as usual and then find the value of c by substituting the coordinates of a point on the line into the equation $y = mx + c$. Alternatively, you can use the coordinates of two points on the line to form and solve a pair of simultaneous equations in m and c. p89

5 To test a belief that the relation between x and y is of the form $y = ax^2 + b$, you need to plot y against x^2. If the points are roughly in a straight line, you can deduce that the relation between x and y is of the form $y = ax^2 + b$. The gradient of the line gives an estimate for a and the intercept on the vertical axis ($X = 0$) gives an estimate for b. p92

6 To test a belief that the relation between x and y is of the form $\frac{1}{x} + \frac{1}{y} = a$, you need to plot $\frac{1}{y}$ against $\frac{1}{x}$. If the points are roughly in a straight line with gradient -1, you can deduce that the relation between x and y is of the form $\frac{1}{x} + \frac{1}{y} = a$. The intercept on the vertical axis ($X = 0$) gives an estimate for a. p94

7 To test a belief that the relation between x and y is of the form $y = ax^2 + bx$, you can plot $\frac{y}{x}$ against x. If the points are roughly in a straight line, you can deduce that the relation between x and y is of the form $y = ax^2 + bx$. The gradient of the line gives an estimate for a and the intercept on the vertical axis ($X = 0$) gives an estimate for b. p96

8 Taking logarithms of both sides of $y = ax^n \Rightarrow \log y = \log a + n \log x$. p100

9 To test a belief that the relation between x and y is of the form $y = ax^n$, you need to plot $\log y$ against $\log x$. If the points are roughly in a straight line, you can deduce that the relation between x and y is of the form $y = ax^n$. The gradient of the line gives an estimate for n and the intercept on the vertical axis ($X = 0$) gives the value of $\log a$ from which the estimate for a can be found. p100

10 Taking logarithms of both sides of $y = ab^x \Rightarrow \log y = \log a + x \log b$. p104

11 To represent the relation $y = ab^x$ in a linear form you need to plot $\log y$ against x. If a straight line is obtained from the given data the relation is true. The gradient of the line is the value of $\log b$ and the intercept on the vertical axis ($X = 0$) gives the value of $\log a$. Knowing these two values, estimates for a and b can be found. p104

Test yourself

Test yourself	What to review
1 The diagram shows a straight line graph of x^2y against $\sqrt{x}$ passing through the points (2, 4) and (6, 2). Express y in the form $ax^p + bx^q$, where a, b, p and q are constants.	Section 7.2
2 The diagram shows a straight line graph of $\log_{10}y$ against $\log_{10}x$, passing through the points (0, 1.2) and (4, 4). Express y in the form ax^b, where a, and b are constants to be found.	Section 7.3
3 Given that $P = ab^t$, where a and b are constants, express $\log P$ in terms of $\log a$, $\log b$ and t. Hence, given that $P = 800$ when $t = 3$ and $P = 1000$ when $t = 4$, find, to three significant figures, the value of a and the value of b.	Section 7.4

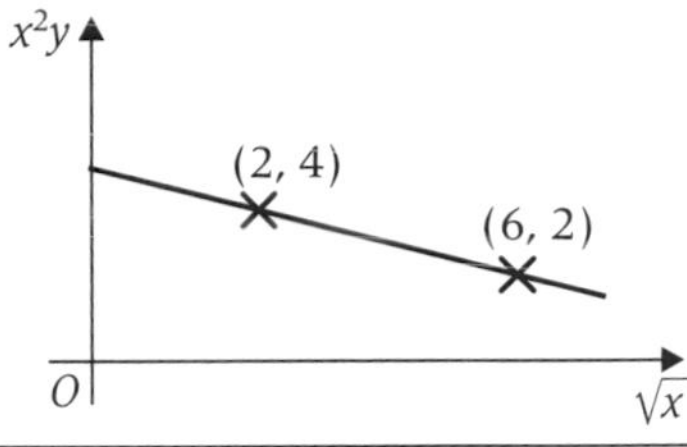

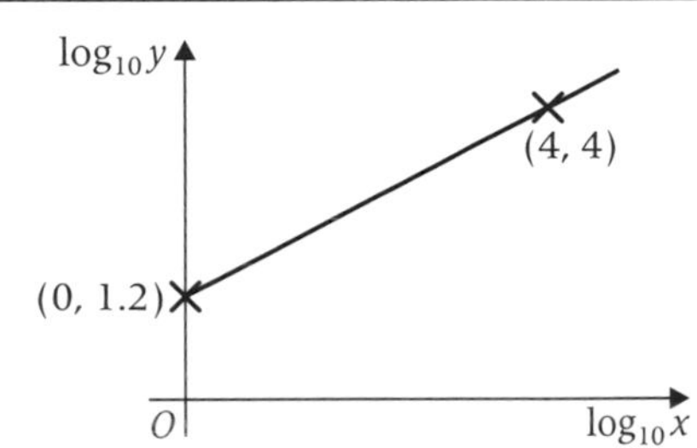

7

Test yourself ANSWERS

1 $y = -\frac{1}{2}x^{-\frac{3}{2}} + 5x^{-2}$.

2 $y = 15.8489\ldots x^{0.7}$

3 $\log P = \log a + t \log b$, $a = 410$ (3 s.f.), $b = 1.25$

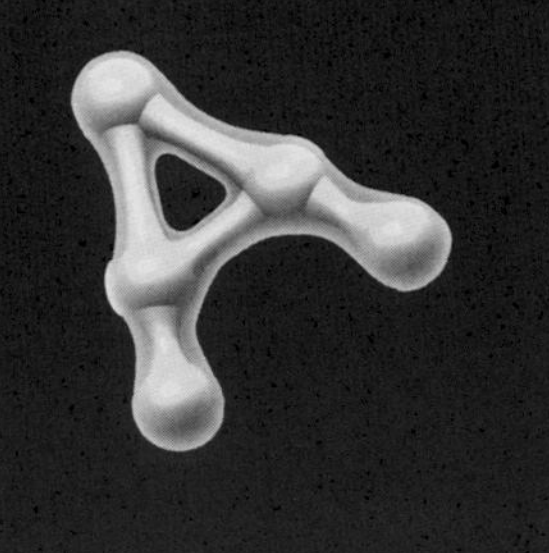

CHAPTER 8

Calculus

Learning objectives

After studying this chapter, you should be able to:

- find the gradient of a chord whose x-coordinates differ by h
- find the gradient of a tangent by considering the limit of the gradient of a chord
- understand what is meant by an improper integral
- evaluate simple improper integrals.

8.1 The gradient of a chord and the gradient of a tangent

You may wish to review sections 9.1 to 9.5 of chapter 9 of C1 as an introduction to the material which follows.

Suppose P and Q are two points on a curve with equation $y = \mathrm{f}(x)$. If you are given the x-coordinates of P and Q, you can substitute into the equation of the curve to find the corresponding y-coordinates.

Let the points have coordinates $P(x_P, y_P)$ and $Q(x_Q, y_Q)$.

The gradient of the chord PQ is $\dfrac{y_Q - y_P}{x_Q - x_P}$.

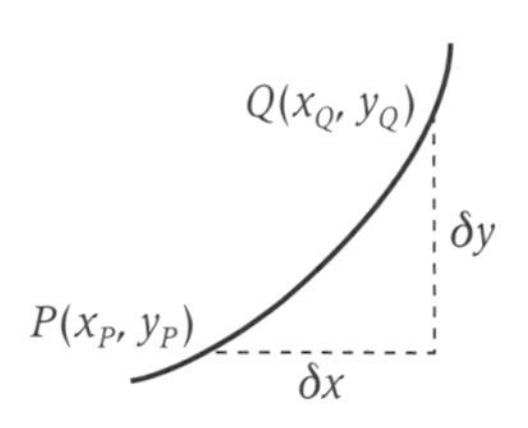

You will be mostly concerned with the case when the x-coordinates differ by a small amount h or δx.

You can then write

$$\delta x = x_Q - x_P \quad \text{and}$$
$$\delta y = y_Q - y_P,$$

and hence the gradient of the chord is $\dfrac{\delta y}{\delta x}$.

As h (or δx) gets smaller and smaller, we say that h tends to zero and write $h \to 0$.

We also say that $\dfrac{\delta y}{\delta x} \to \dfrac{\mathrm{d}y}{\mathrm{d}x}$ and thus obtain the gradient of the curve at the point P.

Worked example 8.1

A curve has equation $y = x^3 + 1$.

Find the gradient of the chord joining the following points:

(a) $x = 2$ and $x = 2.01$,

(b) $x = 2$ and $x = 2 + h$.

Use your answers to **(a)** and **(b)** to suggest what the gradient of the tangent to the curve is when $x = 2$.

Solution

(a) When $x = 2$, $y = 9$ and when $x = 2.01$, $y = 9.120\,601$.

Hence, $\delta x = 2.01 - 2 = 0.01$

and $\delta y = 9.120\,601 - 9 = 0.120\,601$.

The gradient of the chord $= \dfrac{\delta y}{\delta x} = \dfrac{0.120\,601}{0.01} = 12.0601$.

(b) When $x = 2$, $y = 9$ and when $x = 2 + h$, $y = 1 + (2 + h)^3$.

$(2 + h)^3 = 2^3 + 3(2)^2h + 3(2)\,h^2 + h^3 = 8 + 12h + 6h^2 + h^3$

Hence, $\delta x = 2 + h - 2 = h$

and $\delta y = (9 + 12h + 6h^2 + h^3) - 9 = 12h + 6h^2 + h^3$.

The gradient of the chord $= \dfrac{\delta y}{\delta x} = \dfrac{12h + 6h^2 + h^3}{h}$

$= 12 + 6h + h^2$.

As h tends to zero in the answer to **(b)**, the gradient of the chord tends to 12. The answer to **(a)** also suggests that the gradient of the tangent to the curve is 12 at the point where $x = 2$.

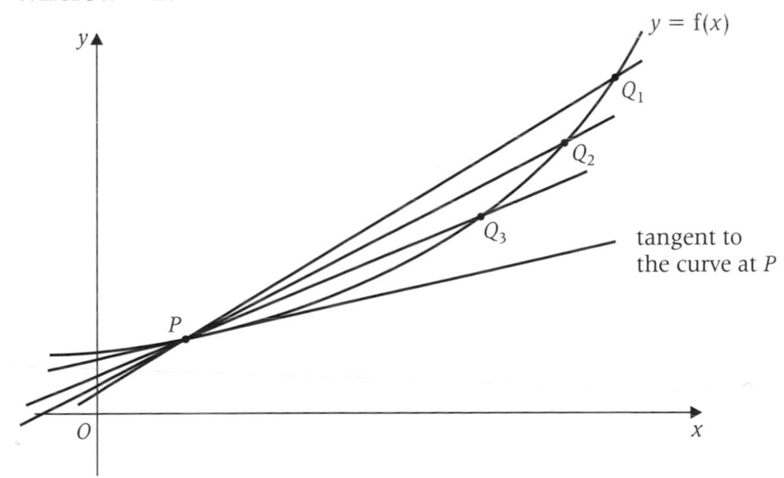

Consider a point P on the curve with equation $y = f(x)$. The position of P is fixed but a second point Q moves along the curve so that it gets closer and closer to P.
Assume that Q starts at position Q_1 then moves to position Q_2, then to Q_3, etc.

When the two points are infinitely close together, the resulting chord becomes the **tangent to the curve at *P***.

This forms the basis of a method for finding the gradient of a curve at any point. It is sometimes called differentiation from first principles.

8.2 The gradient of a curve at a point as the limit of the gradient of the chord

Worked example 8.2

A curve has equation $y = x^2 - 3x + 7$ and the points P and Q have x-coordinates 2 and $2 + h$, respectively.

(a) Find the y-coordinate of Q.

(b) Show that the gradient of the chord PQ is $1 + 4h$.

(c) Deduce the gradient of the curve at P.

Solution

(a) $y_Q = (2 + h)^2 - 3(2 + h) + 7 = 4 + 4h + h^2 - 6 - 3h + 7$
$= 5 + h + h^2$

The diagram shows the situation:

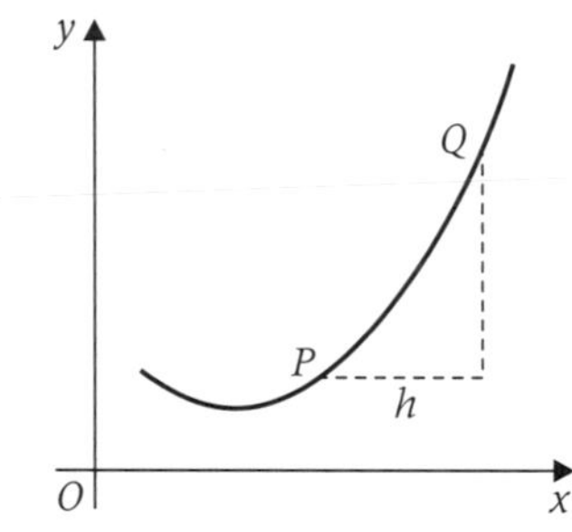

(b) P has coordinates $(2, 5)$

$$\Rightarrow \quad \text{gradient of chord } PQ = \frac{(5 + h + h^2) - 5}{(2 + h) - 2}$$

$$= \frac{h + h^2}{h}$$

$$= \frac{h(1 + h)}{h}$$

h is a common factor in the numerator.

$$= 1 + h.$$

h has been cancelled in the top and bottom.

$$\Rightarrow \quad \text{gradient of } PQ = 1 + h.$$

(c) The gradient of the curve at P = gradient of PQ as $h \to 0$.

The gradient of the curve at $P = \lim_{h \to 0} (1 + h)$

The limit as *h* tends to zero.

Therefore, the gradient of the curve at P is $1 + 0 = 1$.

EXERCISE 8A

1 A curve has equation $y = x^2$. The points P and Q on the curve have coordinates $(3, 9)$ and $(3 + h, (3 + h)^2)$, respectively.

(a) Find the gradient of the chord PQ in as simplified a form as possible.

(b) Hence find the gradient of the curve at the point P.

2 A curve has equation $y = 3x^2 + 2$. The points S and T on the curve have coordinates $(1, 5)$ and $(1 + h, 3(1 + h)^2 + 2)$, respectively.

(a) Show that the gradient of the chord ST is $6 + 3h$.

(b) Hence find the gradient of the curve at the point S.

3 The points P and Q on the curve with equation $y = x^2 - 2x + 3$ have coordinates $(2, 3)$ and $(2 + h, (2 + h)^2 - 2(2 + h) + 3)$, respectively.

(a) Show that the y-coordinate of Q is $h^2 + 2h + 3$.

(b) Find the gradient of the chord PQ in as simplified a form as possible.

(c) Hence find the gradient of the curve at the point P.

4 A curve has equation $y = 5x - x^2$. The points A and B on the curve have x-coordinates 2 and $2 + h$, respectively.

(a) Find the gradient of the chord AB in as simplified a form as possible.

(b) Hence find the gradient of the curve at the point A.

5 A curve has equation $y = x^2 - x + 4$. The points P and Q on the curve have x-coordinates -1 and $-1 + h$, respectively.

(a) Find the gradient of the chord PQ in as simplified a form as possible.

(b) Hence find the gradient of the curve at the point P.

6 A curve has equation $y = 3x^2 - 7x - 20$. The points F and G on the curve have x-coordinates -2 and $-2 + h$, respectively.

(a) Find the gradient of the chord FG in as simplified a form as possible.

(b) Hence find the gradient of the curve at the point F.

7 A curve has equation $y = 10x - 2x^2$. The points V and W on the curve have x-coordinates 0.5 and $0.5 + h$, respectively.

(a) Find the gradient of the chord VW in as simplified a form as possible.

(b) Hence find the gradient of the curve at the point V.

8.3 The use of the binomial theorem

It is sometimes necessary to use the binomial theorem in order to expand expressions such as $(3 + h)^5$.

Worked example 8.3

A curve has equation $y = 2x^5$. The points A and B on the curve have x-coordinates 3 and $3 + h$, respectively.

(a) Find the y-coordinate of B in expanded form.

(b) Hence find the gradient of AB in its simplest form.

(c) Deduce the gradient of the curve at the point A.

Solution

(a) The y-coordinate of B is $2(3 + h)^5$.

You need to use the binomial expansion of $(3 + h)^5$:

$$(3 + h)^5 = 3^5 + 5 \times 3^4 \times h + 10 \times 3^3 \times h^2 + 10 \times 3^2 \times h^3 + 5 \times 3 \times h^4 + h^5$$
$$= 243 + 405h + 270h^2 + 90h^3 + 15h^4 + h^5$$

Hence, the y-coordinate of B

$= 2(3 + h)^5 = 486 + 810h + 540h^2 + 180h^3 + 30h^4 + 2h^5$

Recall Pascal's triangle:

1
1 1
1 2 1
1 3 3 1
1 4 6 4 1
1 5 10 10 5 1

(b) The y-coordinate of A is $2(3)^5 = 486$.

Gradient AB

$$= \frac{(486 + 810h + 540h^2 + 180h^3 + 30h^4 + 2h^5) - 486}{(3 + h) - 3}$$

$$= 810 + 540h + 180h^2 + 30h^3 + 2h^4$$

(c) Letting $h \to 0$ gives the gradient of the curve at A.

Hence, gradient at $A = 810$.

EXERCISE 8B

1 A curve has equation $y = x^3$. The points P and Q on the curve have coordinates $(2, 8)$ and $(2 + h, (2 + h)^3)$, respectively.

(a) Find the gradient of the chord PQ in as simplified a form as possible.

(b) Hence find the gradient of the curve at the point P.

2 The points C and D on the curve with equation $y = 4x^3 + 7$ have coordinates $(1, 11)$ and $(1 + h, 4(1 + h)^3 + 7)$, respectively.

(a) Show that the gradient of the chord CD is $12 + 12h + 4h^2$.

(b) Hence find the gradient of the curve at the point C.

3 The points P and Q on the curve with equation $y = x^4 - 5x$ have coordinates $(2, 6)$ and $(2 + h, (2 + h)^4 - 5(2 + h))$, respectively.

(a) Show that the y-coordinate of Q is $h^4 + 8h^3 + 24h^2 + 27h + 6$.

(b) Find the gradient of the chord PQ in as simplified a form as possible.

(c) Hence find the gradient of the curve at the point P.

4 A curve has equation $y = 7x^2 - x^3$. The points A and B on the curve have x-coordinates 3 and $3 + h$, respectively.

(a) Find the gradient of the chord AB in as simplified a form as possible.

(b) Hence find the gradient of the curve at the point A.

5 A curve has equation $y = x^4 - x^3 + 2$. The points P and Q on the curve have x-coordinates -1 and $-1 + h$, respectively.

(a) Find the gradient of the chord PQ in as simplified a form as possible.

(b) Hence find the gradient of the curve at the point P.

6 The points R and S on the curve with equation $y = 4x^5$ have x-coordinates 1 and $1 + h$, respectively.

(a) Find the gradient of the chord RS in as simplified a form as possible.

(b) Hence find the gradient of the curve at the point R.

7 A curve has equation $y = 2x^4 + 5x^3 - x - 1$. The points P and Q on the curve have x-coordinates 2 and $2 + h$, respectively.

(a) Find the gradient of the chord PQ in as simplified a form as possible.

(b) Hence find the gradient of the curve at the point P.

8 The points A and B on the curve with equation $y = 3x^6$ have x-coordinates -1 and $-1 + h$, respectively.

(a) Find the gradient of the chord AB in as simplified a form as possible.

(b) Hence find the gradient of the curve at the point A.

8.4 Improper integrals with limits involving infinity

Sometimes the upper limit of an integral is ∞ or the lower limit is $-\infty$, and so it is necessary to find a limit in order to evaluate the integral. When the limit exists we can find the value of the integral. It is an example of what is called an improper integral.

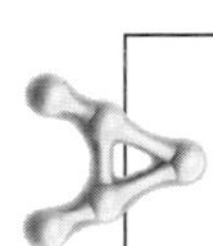

When $f(x)$ is defined for $x \geqslant a$, we define the improper integral $\int_a^{\infty} f(x)\,dx = \lim_{b \to \infty} \int_a^b f(x)\,dx$, provided the limit exists.

Worked example 8.4

(a) Find: **(i)** $\int_1^a x^2\,dx$, and **(ii)** $\int_1^b x^{-2}\,dx$.

(b) Hence determine whether the following integrals exist. If they exist, find their value:

(i) $\int_1^{\infty} x^2\,dx$, and **(ii)** $\int_1^{\infty} x^{-2}\,dx$.

Solution

(a) (i) $\int_1^a x^2\,dx = \left[\frac{x^3}{3}\right]_1^a = \frac{a^3}{3} - \frac{1}{3}$

(ii) $\int_1^b x^{-2}\,dx = \left[\frac{x^{-1}}{-1}\right]_1^b = \frac{b^{-1}}{-1} - \frac{1^{-1}}{-1} = 1 - \frac{1}{b}$

(b) (i) As $a \to \infty$, $a^3 \to \infty$.

Therefore the integral $\int_1^\infty x^2\,dx$ does not exist.

The improper integral $\int_1^\infty x^2\,dx$ is said to diverge.

(ii) As $b \to \infty$, $\frac{1}{b} \to 0$.

You should **not** write $\frac{1}{\infty} = 0$ since ∞ is not a number.

Hence, $\int_1^\infty x^{-2}\,dx = 1 - 0 = 1.$

The integral does exist and has value 1.

The improper integral $\int_1^\infty x^{-2}\,dx$ is sometimes said to converge.

In this case it converges to the value 1.

The improper integral $\int_1^\infty x^{-2}\,dx$ has value 1.

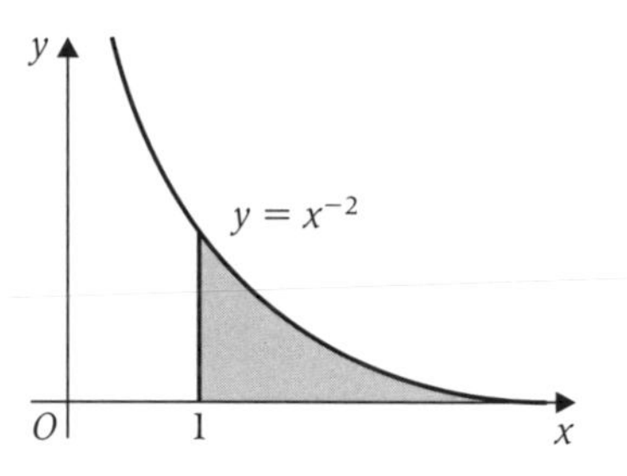

The interesting situation here is that you have an unbounded or infinite region with a finite area.

> When $f(x)$ is defined for $x \leq b$, we define the improper integral $\int_{-\infty}^b f(x)\,dx = \lim_{a \to -\infty} \int_a^b f(x)\,dx$, provided the limit exists.

Worked example 8.5

(a) Find $\int_p^{-2} \frac{1}{x^3}\,dx$.

(b) Hence determine whether the integral $\int_{-\infty}^{-2} \frac{1}{x^3}\,dx$ exists and if it does exist, find its value.

Solution

(a) $\int_p^{-2} x^{-3}\,dx = \left[\frac{x^{-2}}{-2}\right]_p^{-2} = \frac{(-2)^{-2}}{-2} - \frac{p^{-2}}{-2}$

$$= \frac{1}{-2 \times 4} + \frac{p^{-2}}{2} = \frac{1}{2p^2} - \frac{1}{8}$$

(b) As $p \to -\infty$, $\frac{1}{p^2} \to 0$.

Hence, $\int_{-\infty}^{-2} \frac{1}{x^3}\,dx = 0 - \frac{1}{8} = -\frac{1}{8}.$

The improper integral exists and has value $-\frac{1}{8}$.

8.5 Further improper integrals

Sometimes the integrand, $f(x)$ is defined over the interval $p < x < q$ but becomes infinite at one of the end points of the interval.

For example, $f(x) = \frac{1}{\sqrt{x}}$ is defined for $x > 0$, but not when $x = 0$.

$\int_0^4 \frac{1}{\sqrt{x}}\,dx$ is a further type of improper integral and you need to use limits in order to see whether the integral converges or not.

Worked example 8.6

(a) Find $\int_a^4 \frac{1}{\sqrt{x}}\,dx$, where $a > 0$.

(b) Hence determine whether the improper integral $\int_0^4 \frac{1}{\sqrt{x}}\,dx$ exists. If so, find its value.

Solution

(a) $\int_a^4 \frac{1}{\sqrt{x}}\,dx = \int_a^4 x^{-\frac{1}{2}}\,dx$

$= \left[2x^{\frac{1}{2}}\right]_a^4 = 2 \times 4^{\frac{1}{2}} - 2a^{\frac{1}{2}} = 4 - 2\sqrt{a}$

(b) In order to consider whether the integral with lower limit zero exists, it is necessary to let a tend to zero through positive values and we write $a \to 0+$.

As $a \to 0+$, $\quad 4 - 2\sqrt{a} \to 4 - 0 = 4$.

Hence the improper integral $\int_0^4 \frac{1}{\sqrt{x}}\,dx$ exists and its value is 4.

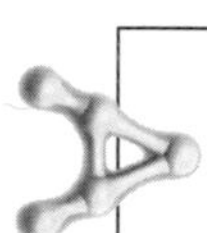

When $f(x)$ is defined for $p < x < q$, but $f(x)$ is not defined when $x = p$, then the improper integral

$\int_p^q f(x)\,dx = \lim_{a \to p+} \int_a^q f(x)\,dx$, provided the limit exists.

EXERCISE 8C

1 (a) Find each of the following integrals:

(i) $\int_2^a x^{-2}\,dx$, **(ii)** $\int_1^b x^{-4}\,dx$, **(iii)** $\int_{\frac{1}{2}}^c x^{-2}\,dx$, **(iv)** $\int_1^d x\,dx$.

(b) Hence determine whether each of the following improper integrals exists. If it exists, find its value.

(i) $\int_2^\infty x^{-2}\,dx$, **(ii)** $\int_1^\infty x^{-4}\,dx$, **(iii)** $\int_{\frac{1}{2}}^\infty x^{-2}\,dx$, **(iv)** $\int_1^\infty x\,dx$.

2 (a) Find each of the following integrals:

(i) $\int_a^{-3} x^{-2}\,dx$, **(ii)** $\int_b^{-1} x^{-5}\,dx$,

(iii) $\int_c^{-1} x^3\,dx$, **(iv)** $\int_d^{-0.5} x^{-6}\,dx$.

(b) Hence determine whether each of the following improper integrals exists. If it exists, find its value.

(i) $\int_{-\infty}^{-3} x^{-2}\,dx$, **(ii)** $\int_{-\infty}^{-1} x^{-5}\,dx$,

(iii) $\int_{-\infty}^{-1} x^3\,dx$, **(iv)** $\int_{-\infty}^{-0.5} x^{-6}\,dx$.

3 (a) Find each of the following integrals:

(i) $\int_a^9 \frac{1}{\sqrt{x}}\,dx$, **(ii)** $\int_b^2 x^{-2}\,dx$,

(iii) $\int_c^8 x^{-\frac{1}{3}}\,dx$, **(iv)** $\int_d^{16} x^{-\frac{1}{4}}\,dx$.

(b) Hence determine whether each of the following improper integrals exists. If it exists, find its value.

(i) $\int_0^9 \frac{1}{\sqrt{x}}\,dx$, **(ii)** $\int_0^2 x^{-2}\,dx$,

(iii) $\int_0^8 x^{-\frac{1}{3}}\,dx$, **(iv)** $\int_0^{16} x^{-\frac{1}{4}}\,dx$.

4 Explain why $\int_0^{81} \frac{1}{\sqrt{x}}\,dx$ is an improper integral and find its value.

5 A student evaluates $\int_{-2}^{a} x^{-2}\,dx$ as $-\frac{1}{a} - \frac{1}{2}$ and concludes that $\int_{-2}^{\infty} x^{-2}\,dx$ is equal to $-\frac{1}{2}$. Explain why she is incorrect.

Key point summary

1 The gradient of the chord PQ is $\frac{y_Q - y_P}{x_Q - x_P}$. *p112*

2 When the x-coordinate of P is a, and the x-coordinate of Q is $a + h$, the gradient of the chord PQ can be simplified to an expression involving h.
The gradient of the curve at the point P is obtained by letting h tend to zero. *p114*

3 When $f(x)$ is defined for $x \geqslant a$, we define the improper integral $\int_a^{\infty} f(x)\,dx = \lim_{b \to \infty} \int_a^b f(x)\,dx$, provided the limit exists. *p117*

4 When $f(x)$ is defined for $x \leqslant b$, we define the improper integral $\int_{-\infty}^{a} f(x)\,dx = \lim_{a \to -\infty} \int_{a}^{b} f(x)\,dx$, provided the limit exists. p118

5 When $f(x)$ is defined for $p < x < q$, but $f(x)$ is not defined when $x = p$, then the improper integral $\int_{p}^{q} f(x)\,dx = \lim_{a \to p+} \int_{a}^{q} f(x)\,dx$, provided the limit exists. p119

Test yourself — What to review

1 A curve has equation $y = 3x - x^2$.
The points P and Q lie on the curve and P has x-coordinate equal to 2 and Q has x-coordinate equal to $2 + h$.
(a) Find the gradient of PQ in its simplest form.
(b) Deduce the gradient of the curve at the point P.
Section 8.2

2 Find the gradient of the chord AB, where A is the point $(3, 81)$ and B is the point $(3 + h, (3 + h)^4)$ on the curve with equation $y = x^4$.
(a) Find the gradient of AB in its simplest form.
(b) Deduce the gradient of the curve at the point B.
Section 8.3

3 **(a)** Find: **(i)** $\int_{1}^{a} x^{-4}\,dx$, **(ii)** $\int_{b}^{-4} x^{-2}\,dx$, **(iii)** $\int_{1}^{c} x^{-\frac{1}{2}}\,dx$.
(b) Hence determine whether each of the following integrals exists. If it exists, find its value.
(i) $\int_{1}^{\infty} x^{-4}\,dx$, **(ii)** $\int_{-\infty}^{-4} x^{-2}\,dx$, **(iii)** $\int_{1}^{\infty} x^{-\frac{1}{2}}\,dx$.
Section 8.4

4 Explain why $\int_{0}^{16} \frac{1}{\sqrt{x}}\,dx$ is an improper integral and find its value. *Section 8.5*

Test yourself ANSWERS

1 **(a)** $\frac{(2 - h - h^2) - 2}{h} = -1 - h$ (in its simplest form); **(b)** -1.

2 **(a)** $h^3 + 12h^2 + 54h + 108$; **(b)** 108.

3 **(a)** **(i)** $\frac{1}{3} - \frac{1}{3a^3}$, **(ii)** $\frac{1}{b} + \frac{1}{4}$, **(iii)** $2\sqrt{c} - 2$;
(b) **(i)** $\frac{1}{3}$, **(ii)** $\frac{1}{4}$, **(iii)** does not exist.

4 The integrand $\frac{1}{\sqrt{x}}$ is not defined when $x = 0$, 8.

CHAPTER 9

Series

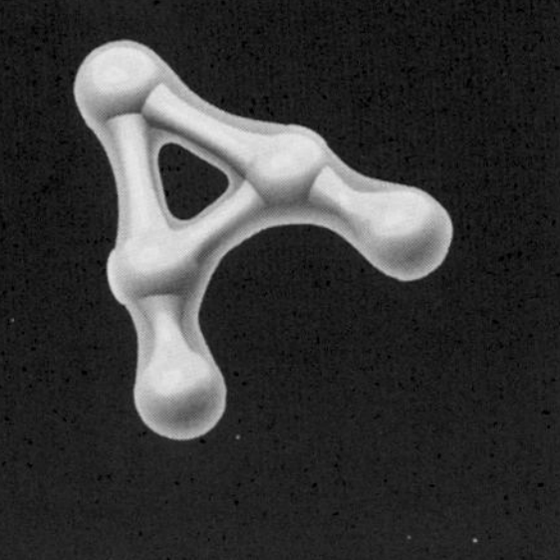

Learning objectives

After studying this chapter, you should be able to:

- use sigma notation
- find Σk, where k is a constant
- find and use a formula for Σr
- use the formulae for Σr^2 and Σr^3 to find sums of other series.

You have met arithmetic series and geometric series in C2. The next section is a short review of the sigma notation used for series.

9.1 Sigma notation

A shorthand notation for finding sums is to use the Greek capital letter sigma, Σ, which corresponds to our letter S.

$$\sum_{k=2}^{10} k^3$$

At the top of the sigma sign is the final value the variable can take.

Shorthand for $2^3 + 3^3 + 4^3 + \ldots + 10^3$.

At the bottom of the sigma sign is the variable which serves as a counter and the initial value it takes.

Worked example 9.1

Write each of the following in full and hence find the value of:

(a) $\sum_{k=4}^{8} k^2$ **(b)** $\sum_{r=2}^{5} (2^r + 1)$, **(c)** $\sum_{n=2}^{5} (n^3 - 8)$.

Note that different letters may be used *within* the summation such as k, r, n, etc. and yet the final answer will not involve any of these variables.

Solution

(a) The expression to be summed here is k^2.

You need to substitute $k = 4$, then $k = 5$ … up to $k = 8$ adding the terms together.

$$\sum_{k=4}^{8} k^2 = 4^2 + 5^2 + 6^2 + 7^2 + 8^2 = 190$$

(b) This time the summation expression is $(2^r + 1)$ with r ranging from 2 to 5.

$$\sum_{r=2}^{5} (2^r + 1) = 5 + 9 + 17 + 33$$
$$= 64$$

(c) Substituting $n = 2, 3, 4, 5$ and summing gives

$$\sum_{n=2}^{5} (n^3 - 8) = 0 + 19 + 56 + 117 = 192.$$

EXERCISE 9A

Write each of the following in full and hence find the value of:

1 $\sum_{k=2}^{6} k$

2 $\sum_{r=6}^{10} (2r - 5)$

3 $\sum_{n=2}^{5} (n^2 + 1)$

4 $\sum_{r=6}^{10} (4r - 29)$

5 $\sum_{k=2}^{5} k^3$

6 $\sum_{n=3}^{7} (n - 2)(n - 3)$

7 $\sum_{r=1}^{7} (r - 5)(r - 4)(r - 3)$

8 $\sum_{n=10}^{14} (n - 12)(n - 11)(n - 10)$

9 $\sum_{k=0}^{5} (k^3 - k^2)$

10 $\sum_{r=4}^{10} (r - 7)(r - 8)(r - 9)$

9.2 The sum of the first *n* natural numbers

You may recall in C2 chapter 8 that you learnt how to find the sum

$$S = 1 + 2 + 3 + 4 + 5 + \ldots + n.$$

By writing the terms in reverse order underneath the terms of the first series, you obtain

$$S = 1 + 2 + 3 + \ldots + (n - 1) + n$$
$$S = n + (n - 1) + (n - 2) + \ldots + 2 + 1$$

which on adding gives n quantities each with value $n + 1$.

Therefore $2S = n(n + 1)$. Hence, $S = \frac{1}{2}n(n + 1)$.

Writing this in sigma notation gives:

$$1 + 2 + 3 + 4 + \ldots + n = \sum_{r=1}^{n} r = \frac{1}{2}n(n + 1)$$

Of course any letter (other than n) could be used instead of r in the summation.

Another important result is that

$$\sum_{r=1}^{n} 1 = \underbrace{1 + 1 + 1 + \ldots + 1}_{n \text{ terms}} = n$$

This follows because you are adding the number 1 'n times'.

It is often helpful to split a series up into separate summations:

$$\sum \left[A\mathrm{f}(r) + B\mathrm{g}(r)\right] = A\sum \mathrm{f}(r) + B\sum \mathrm{g}(r)$$

9

Worked example 9.2

Find the value of:

(a) $\sum_{r=1}^{50}(r+3)$, **(b)** $\sum_{k=1}^{200}(3k-100)$.

Solution

(a) You can split the summation into two parts:

$$\begin{aligned}\sum_{r=1}^{50}(r+3) &= \sum_{r=1}^{50} r + \sum_{r=1}^{50} 3\\ &= \left(\tfrac{1}{2}\times 50\times 51\right) + (3\times 50)\\ &= 1275 + 150\\ &= 1425\end{aligned}$$

Using the results in the boxes on the previous page.

(b) This time you can take the constant 3 outside the summation:

$$\begin{aligned}\sum_{k=1}^{200}(3k-100) &= 3\sum_{k=1}^{200} k - \sum_{k=1}^{200} 100\\ &= 3\left(\tfrac{1}{2}\times 200\times 201\right) - (100\times 200)\\ &= 60\,300 - 20\,000\\ &= 40\,300\end{aligned}$$

Worked example 9.3

Find the value of $\sum_{r=20}^{80}(4r-7)$.

Two methods are shown since the techniques can be used in the more difficult sections that follow.

Solution

Method 1

You can rewrite the summation as the difference of two series:

$$\sum_{r=20}^{80}(4r-7) = \sum_{r=1}^{80}(4r-7) - \sum_{r=1}^{19}(4r-7)$$

$$\begin{aligned}\sum_{r=1}^{80}(4r-7) &= 4\sum_{r=1}^{80} r - (7\times 80) = \left(\tfrac{4}{2}\times 80\times 81\right) - 560\\ &= 12\,960 - 560 = 12\,400\end{aligned}$$

$$\begin{aligned}\sum_{r=1}^{19}(4r-7) &= 4\sum_{r=1}^{19} r - (7\times 19) = \left(\tfrac{4}{2}\times 19\times 20\right) - 133\\ &= 760 - 133 = 627\end{aligned}$$

Therefore $\sum_{r=20}^{80}(4r-7) = 12\,400 - 627 = 11\,773$

Method 2
You can make a substitution so you have a new summation variable starting at 1.

In this case, let $k = r - 19$ so that r is replaced by $k + 19$. The counter for k goes from 1 to 61.

$$\sum_{r=20}^{80} (4r - 7) = \sum_{k=1}^{61} [4(k + 19) - 7] = \sum_{k=1}^{61} (4k + 69)$$

$$= 4\sum_{k=1}^{61} k + (61 \times 69)$$

$$= \tfrac{4}{2} \times 61 \times 62 \times 4209$$

$$= 7564 + 4209$$

$$= 11\,773$$

In C2 you used arithmetic series. This is an arithmetic series with first term 73 and common difference 4. The sum of 61 terms is $\frac{61}{2} \times (146 + 60 \times 4) = 11\,773$.

9.3 The sums of squares and cubes

You have already seen that $1 + 2 + 3 + 4 + \ldots + n = \frac{1}{2}n(n + 1)$ and that this can be written in sigma notation as

$$\sum_{r=1}^{n} r = \tfrac{1}{2}n(n + 1).$$

There are similar formulae for the sums of the squares and cubes of the first n natural numbers. The formulae will be proved in a later module, but you only need to be able to **use** these formulae in the FP1 examination.

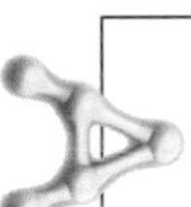

Sums of squares

$$1^2 + 2^2 + 3^2 + 4^2 + \ldots + n^2 = \sum_{r=1}^{n} r^2 = \tfrac{1}{6}n(n + 1)(2n + 1)$$

Sums of cubes

$$1^3 + 2^3 + 3^3 + 4^3 + \ldots + n^3 = \sum_{r=1}^{n} r^3 = \tfrac{1}{4}n^2(n + 1)^2$$

Worked example 9.4

Evaluate **(a)** $\sum_{r=11}^{50} r^2$, **(b)** $\sum_{k=100}^{200} k^3$.

Solution

(a) $\sum_{r=11}^{50} r^2 = \sum_{r=1}^{50} r^2 - \sum_{r=1}^{10} r^2$

$$= \tfrac{1}{6} \times 50 \times 51 \times 101 - \tfrac{1}{6} \times 10 \times 11 \times 21 \ldots$$

$$= 42\,540$$

Now you can use the formulae above with $n = 50$ and $n = 10$.

(b) $\sum_{k=100}^{200} k^3 = \sum_{k=1}^{200} k^3 - \sum_{k=1}^{99} k^3$

$$= \tfrac{1}{4} \times 200^2 \times 201^2 - \tfrac{1}{4} \times 99^2 \times 100^2$$

$$= 379\,507\,500$$

Sometimes a combination of these formulae needs to be used.

Remember: $\sum_{r=1}^{n} 1 = \underbrace{1 + 1 + 1 + \ldots + 1}_{n \text{ terms}} = n$

A common mistake is to think this is equal to 1.

You can split up a complicated summation into separate sums, for example;

$$\sum(3k^2 + 2k + 5) = \sum 3k^2 + \sum 2k + \sum 5.$$

You can take a constant outside the sigma sign as a common factor so that $\sum 3k^2 + \sum 2k = 3\sum k^2 + 2\sum k$ and then you can use the standard formulae to evaluate the summation.

Worked example 9.5

Find the value of $\sum_{r=1}^{25} (2r^2 - 5r + 4)$.

Solution

$$\sum_{r=1}^{25} (2r^2 - 5r + 4) = \sum_{r=1}^{25} 2r^2 - \sum_{r=1}^{25} 5r + \sum_{r=1}^{25} 4$$

$$= 2\sum_{r=1}^{25} r^2 - 5\sum_{r=1}^{25} r + 4\sum_{r=1}^{25} 1$$

Here, $\sum_{r=1}^{25} r^2 = \frac{1}{6} \times 25 \times 26 \times 51 = 5525$,

$\sum_{r=1}^{25} r = \frac{1}{2} \times 25 \times 26 = 325$,

and $\sum_{r=1}^{25} 1 = 25$

$$\Rightarrow \sum_{r=1}^{25} (2r^2 - 5r + 4) = 2 \times 5525 - 5 \times 325 + 4 \times 25 = 9525$$

EXERCISE 9B

1 Find the value of each of the following:

(a) $\sum_{r=1}^{25} 2$, **(b)** $\sum_{r=1}^{20} r^2$, **(c)** $\sum_{r=1}^{40} r^3$,

(d) $\sum_{r=51}^{100} r^2$, **(e)** $\sum_{r=101}^{130} r^3$, **(f)** $\sum_{r=101}^{140} (r^3 - 500)$.

2 Evaluate:

(a) $\sum_{k=1}^{10} 6k^2$, **(b)** $\sum_{k=1}^{60} 4k^3$, **(c)** $\sum_{k=16}^{30} 3k^2$,

(d) $\sum_{k=15}^{50} (k^3 - 100)$, **(e)** $\sum_{k=11}^{30} (k^2 + 3k)$, **(f)** $\sum_{k=1}^{20} (3k^3 - 6k^2 + 7)$.

3 Find the value of $\sum_{k=1}^{20} (4k^3 - 36k^2 - 10)$.

4 Find the value of $\sum_{k=13}^{50} (2k^3 - 12k^2 - 7)$.

5 Evaluate: **(a)** $\sum_{k=1}^{20} (2k - 3)^2$, **(b)** $\sum_{k=11}^{40} (3k - 2)^2$.

6 Calculate: **(a)** the sum of, and **(b)** the sum of the squares of the integers from 500 to 1000 inclusive.

7 Find the least value of n for which $\sum_{r=1}^{n} r^3 > 10\,000$.

8 **(a)** Find the value of: **(i)** $\sum_{r=1}^{100} r^3$, **(ii)** $\sum_{r=51}^{100} r^3$.

(b) Find the sum of the fifty integers from 51 to 100 inclusive.

(c) Hence find the value of $\sum_{r=51}^{100} (r^3 - 6325r)$. [A]

9.4 Questions involving algebra

The questions so far have all had numerical answers.

Sometimes the questions need you to use algebraic skills, looking for common factors and using other techniques to simplify algebraic expressions.

Worked example 9.6

Find $\sum_{k=1}^{n} (4k^3 - 12k)$ and factorise your answer.

Solution

$$\sum_{k=1}^{n} (4k^3 - 12k) = 4\sum_{k=1}^{n} k^3 - 12\sum_{k=1}^{n} k$$

$$= 4 \times \tfrac{1}{4} \times n^2 \times (n + 1)^2 - 12 \times \tfrac{1}{2} \times n \times (n + 1)$$

$$= n(n + 1)[n(n + 1) - 6]$$

$$= n(n + 1)(n^2 + n - 6)$$

$$= n(n + 1)(n + 3)(n - 2)$$

The simplification is much easier if you spot the common factors.

Finally you factorise the quadratic.

EXERCISE 9C

1 Prove that $\sum_{r=1}^{n} (6r^2 + 24r) = n(n + 1)(2n + 13)$.

2 Prove that $\sum_{r=1}^{n} (4r^3 + 6r) = n(n + 1)(n^2 + n + 3)$.

3 Prove that $\sum_{r=1}^{n} 12r^2(r+1) = n(n+1)(3n^2+7n+2)$.

4 Prove that $\sum_{r=1}^{n} (8r^3+6r-3) = n^2(2n^2+4n+5)$.

5 Prove that $\sum_{r=1}^{n} (2r^3+6r-3) = \frac{1}{2}n^2(n^2+2n+7)$.

6 Obtain an expression in terms of n for $\sum_{k=1}^{n} 2k(10k^2+1)$, giving your answer as a product of factors.

7 Find the sums of each of the following, giving your answers in factorised form:

(a) $\sum_{r=1}^{n} (4r^3+2r)$,

(b) $\sum_{r=1}^{n} (6r^2+4r)$,

(c) $\sum_{r=1}^{n} (8r^3-2r)$,

(d) $\sum_{r=1}^{n} (8r^3-6r^2)$,

(e) $\sum_{r=1}^{n} (12r^3+8r)$,

(f) $\sum_{r=1}^{n} (6r^2-6)$.

Key point summary

1 $\sum_{r=1}^{n} 1 = 1+1+1+\ldots+1 = n$ p123

2 $1+2+3+4+\ldots+n = \sum_{r=1}^{n} r = \frac{1}{2}n(n+1)$ p123

3 $\sum[Af(r)+Bg(r)] = A\sum f(r) + B\sum g(r)$ p123

4 $1^2+2^2+3^2+4^2+\ldots+n^2 = \sum_{r=1}^{n} r^2 = \frac{1}{6}n(n+1)(2n+1)$ p125

5 $1^3+2^3+3^3+4^3+\ldots+n^3 = \sum_{r=1}^{n} r^3 = \frac{1}{4}n^2(n+1)^2$ p125

Formulae for Key Points 2, 4 and 5 are in the formulae book for use in examinations, under the heading Summations.

Test yourself	What to review
1 Write each of the following in full and hence find the value of: **(a)** $\sum_{k=3}^{6} k^2$, **(b)** $\sum_{r=6}^{10} (r-6)(r-7)(r-10)$.	*Section 9.1*
2 Find the value of: **(a)** $\sum_{k=1}^{40} k^2$, **(b)** $\sum_{r=1}^{80} (r^3 - 100\,000)$.	*Section 9.3*
3 Find the value of $\sum_{r=41}^{100} (2r^3 - 3r^2)$.	*Section 9.3*
4 Prove that $\sum_{r=1}^{n} (r^3 - r) = \frac{1}{4} n(n+1)(n+2)(n-1)$.	*Section 9.4*

Test yourself ANSWERS

1 **(a)** 86; **(b)** −10.

2 **(a)** 22 140; **(b)** 2 497 600.

3 48 711 570.

9

CHAPTER 10

Numerical methods

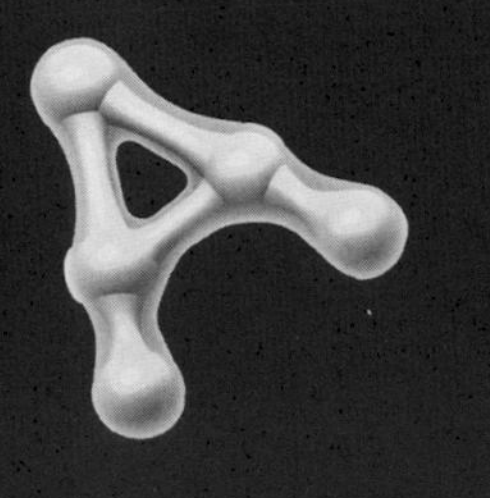

Learning objectives

After studying this chapter, you should be able to:

- recognise how a change in sign of $f(x)$ can be used to find an interval in which a root of $f(x) = 0$ lies
- find approximations to roots of equations by interval bisection
- use the Newton–Raphson method as an iterative method for finding roots of equations
- use a step-by-step method to find linear approximations to solutions of differential equations of the form $\frac{dy}{dx} = f(x)$.

10.1 Change of sign to find roots of equations

Suppose the curve with equation $y = f(x)$ is continuous and crosses the x-axis between $x = a$ and $x = b$.

The situation may be as in diagram 1 or as in diagram 2.

Diagram 1

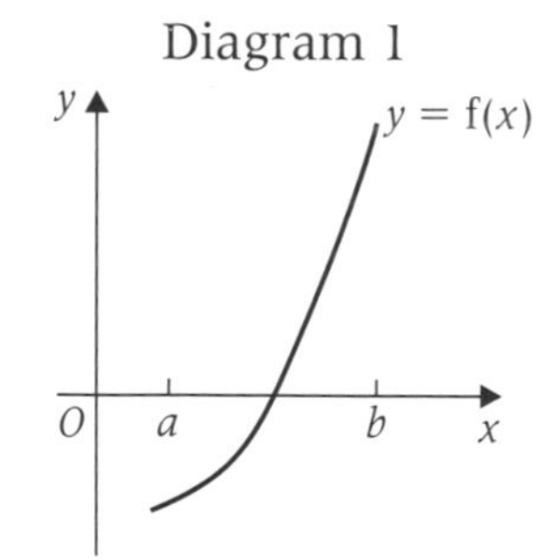

Diagram 2

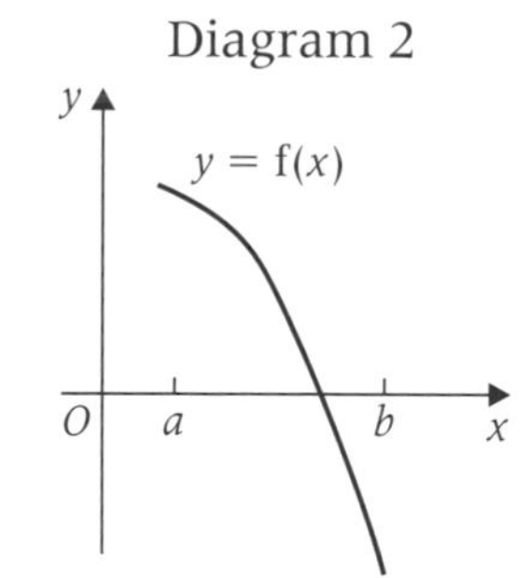

In diagram 1,
$f(a) < 0$ and $f(b) > 0$

In diagram 2,
$f(a) > 0$ and $f(b) < 0$.

In each case there is a change in sign, either going from negative to positive or from positive to negative.

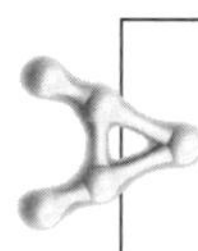

> If the graph of $y = f(x)$ is continuous over the interval $a \leqslant x \leqslant b$, and $f(a)$ and $f(b)$ have different signs, then a root of the equation $f(x) = 0$ must lie in the interval $a < x < b$.

Worked example 10.1

Show that the real root of the equation $x^3 - x + 2 = 0$ lies between -1.6 and -1.5.

You could check this on your graphics calculator by drawing the graph and using the trace facility to see where the curve crosses the x-axis.

Solution

Let $f(x) = x^3 - x + 2$.

The graph of $y = f(x)$ is continuous for all values of x.

$$f(-1.6) = (-1.6)^3 - (-1.6) + 2 = -4.096 + 1.6 + 2 = -0.496$$
$$f(-1.5) = (-1.5)^3 - (-1.5) + 2 = -3.375 + 1.5 + 2 = 0.125$$

The change in sign indicates that the root lies between -1.6 and -1.5.

Worked example 10.2

Find two consecutive integers between which the real root of the equation $x^5 + x = 50$ lies. (You may assume the equation has a single real root.)

Solution

Always set up a function first so the equation is in the form $f(x) = 0$.

Let $f(x) = x^5 + x - 50$.

The graph of $y = f(x)$ is continuous for all values of x.

$$f(0) = 0 + 0 - 50 = -50$$
$$f(2) = 32 + 2 - 50 = -16$$
$$f(3) = 243 + 3 - 50 = 196$$

The root cannot lie between 0 and 2 since both of these values are negative.

There is now a change of sign between these last two values of x.

The two consecutive integers between which the root lies are 2 and 3.

In each of the examples above, the functions considered were polynomials and the graphs of all polynomials are continuous. You cannot apply the change of sign test when the function you are considering is not continuous over the interval concerned.

Worked example 10.3

(a) Given that $f(x) = \frac{1}{x}$. Find $f(2)$ and $f(-1)$.

(b) Can you deduce that a root of $\frac{1}{x} = 0$ lies between -1 and 2?

Solution

(a) $f(2) = \frac{1}{2}$ and $f(-1) = -1$.

(b) Although there is a change in sign, you need to consider the graph of $y = f(x)$.

The graph is **not** continuous between -1 and 2. It has a discontinuity when $x = 0$, so you **cannot** deduce that the equation has a root between -1 and 2.

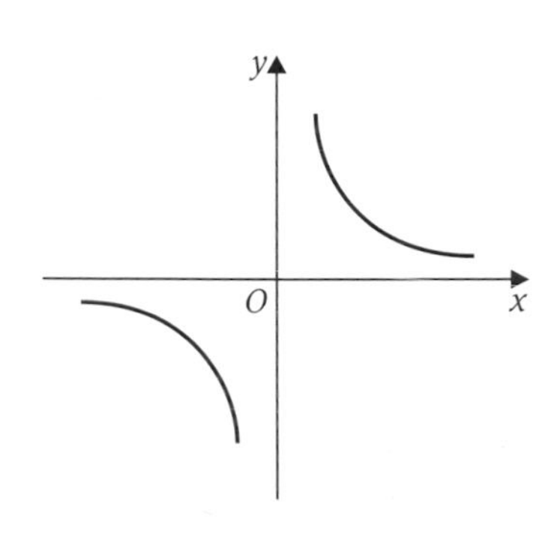

Worked example 10.4

Prove that the real root of the equation $x^3 - 3x + 4 = 0$ is -2.196 correct to three decimal places.

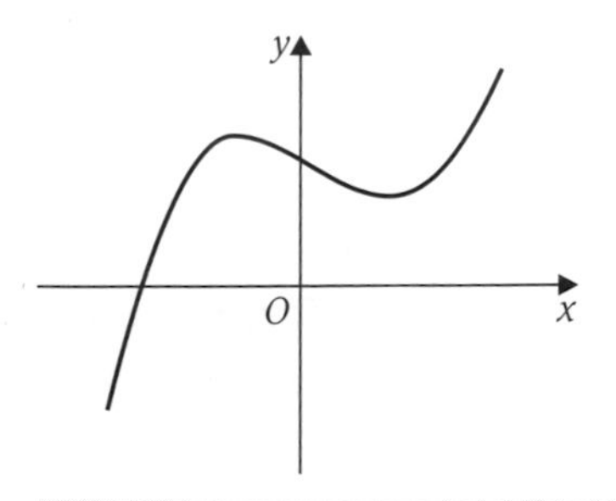

Solution

Your graphics calculator will show that the graph of $y = x^3 - 3x + 4$ is continuous and that it crosses the x-axis once only.

Let $f(x) = x^3 - 3x + 4$.

In order to prove that the root is -2.196 correct to three decimal places, it is necessary to show that the root lies between -2.1965 and -2.1955.

Can you see why?

$f(-2.1965) = -0.007\,760\,8$ and $f(-2.1955) = 0.003\,706\,4\ldots$

The change of sign shows that the root lies between -2.1965 and -2.1955. Hence the root is equal to -2.196 correct to three decimal places.

EXERCISE 10A

1 Verify that the equation $x^3 + 7x - 37 = 0$ has a root between 2.6 and 2.7.

2 Show that the equation $x^3 - 2x + 7 = 0$ has a root between -2.3 and -2.2.

3 Prove that the equation $x^5 + 4x = 9$ has a root between 1.30 and 1.31.

4 Verify that the equation $2^x + x = 30$ has a root with value 4.66 to three significant figures.

5 Prove that one root of the equation $x^4 - 6x + 4 = 0$ lies between 0 and 1 and find two consecutive integers between which the other real root lies.

6 Find two consecutive integers between which the real root of the equation $x^3 + 5x + 8 = 0$ lies.

10.2 The bisection method

A systematic way to make use of this change of sign technique is the bisection method. As the name suggests, this involves repeated bisection of the interval in which the root is known to lie.

Worked example 10.5

(a) Prove that the equation $x^4 + x^3 = 1$ has a root between 0.78 and 0.82.

(b) Use the bisection method to find an interval of width 0.01 in which the root lies.

Solution

(a) Let $f(x) = x^4 + x^3 - 1$.

$$f(0.78) = 0.78^4 + 0.78^3 - 1 = -0.155\ldots$$

$$f(0.82) = 0.82^4 + 0.82^3 - 1 = 0.003\,49\ldots$$

Since the graph of $y = f(x)$ is continuous, the change of sign implies that at least one root lies in the interval $0.78 < x < 0.82$.

(b) Bisecting the interval means you **must** next try $f(0.80)$, even though you expect the root to lie fairly close to 0.82.

$$f(0.80) = 0.80^4 + 0.80^3 - 1 = -0.0784\ldots$$

The change of sign means that root lies in the interval $0.80 < x < 0.82$.

Again, by bisecting the interval, your next evaluation must be $f(0.81)$.

$$f(0.81) = 0.81^4 + 0.81^3 - 1 = -0.038\,09\ldots$$

Hence at least one root lies in the interval $0.81 < x < 0.82$.

This is now an interval of width 0.01 and so there is no need to find the root to any greater precision.

When a root of $f(x) = 0$ is known to lie between $x = a$ and $x = b$, the bisection method requires you to next find the value of $f\left(\frac{a+b}{2}\right)$.

There must then be a change of sign which allows you to bisect the interval in which the root lies.
The procedure is repeated until you have an interval of the desired width containing the root.

Decimal search

Although the bisection method is an easy algorithm to follow, the process of bisecting the interval may lead to an answer such as 'the root lies between 0.818 75 and 0.820 00'.

Since we work with a decimal and not a binary system, this kind of answer is not very user-friendly.

An improvement is called **decimal search**. This involves finding a root correct to one decimal place, then to two decimal places, etc.

This technique involves lots of calculations and is quite tedious and will not be tested in the FP1 unit.

EXERCISE 10B

1 **(a)** Show that the equation $x^3 - 3x + 7 = 0$ has a root between -2.2 and -2.6.

(b) Use the bisection method to find an interval of width 0.1 in which the root lies.

2 (a) Show that the equation $x^3 + x - 5 = 0$ has a root between 1.4 and 1.8.

(b) Use the bisection method to find an interval of width 0.1 in which the root lies.

3 (a) Show that the equation $x^5 + 2x - 18 = 0$ has a root between 1.6 and 1.8.

(b) Use the bisection method to find an interval of width 0.05 in which the root lies.

4 (a) Show that the equation $x^5 + 7x^3 - 2 = 0$ has a root between 0.6 and 0.7.

(b) Use the bisection method to find an interval of width 0.0125 in which the root lies.

5 Sketch the graphs of $y = 2^x$ and $y = 2 - x$ on the same axes. Explain why the equation $2^x + x - 2 = 0$ has exactly one real root α.

(a) Show that α lies between 0.3 and 0.7.

(b) Use the bisection method, showing your working clearly, to find an interval of width 0.1 which contains α. [A]

6 Prove that the equation $\sqrt{(x + 1)} - \sqrt{x} = 0.2$ has a root between 4 and 6. Use the bisection method to find an interval of width 0.125 in which the root lies.

7 (a) Show that the equation $3x - 4\cos x = 0$ has a root between 0.8 and 0.9.

(b) Use the bisection method to find an interval of width 0.025 in which the root lies.

Remember to change your calculator settings to radians in order to answer questions 7 and 8.

8 (a) Show that the equation $2x + 3 - \sin x = 0$ has a root between -2.0 and -1.9.

(b) Use the bisection method to find an interval of width 0.025 in which the root lies.

9 (a) Use logarithms to solve the equation $2^x = 7$, giving your answer to three significant figures.

(b) The equation $2^x = 7 - x$ has a single root α.

(i) Show that α lies between 2.0 and 2.4.

(ii) Use the bisection method to find an interval of width 0.1 in which α lies. [A]

10.3 Linear interpolation

The process of decimal search can be speeded up by the method of linear interpolation.

When a root of $f(x) = 0$ is known to lie between two particular values $x = a$ and $x = b$, the graph of $y = f(x)$ must cross the x-axis at some value between a and b.

The method approximates the curve by a straight line joining the two points and uses the value of $x = c$, where the straight line crosses the x-axis as the next approximation.

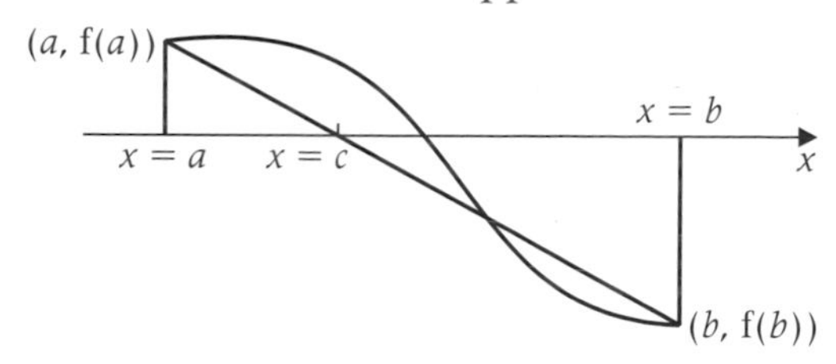

The straight line through the points has equation

$$\frac{y - f(a)}{x - a} = \frac{f(b) - f(a)}{b - a}.$$

When $y = 0$, $x = c$, therefore $c - a = -f(a)\ \dfrac{b - a}{f(b) - f(a)}$.

Hence,

$$\begin{aligned} c &= a - f(a)\ \frac{b - a}{f(b) - f(a)} \\ &= \frac{a[f(b) - f(a)] - f(a)[b - a]}{f(b) - f(a)} \\ &= \frac{af(b) - bf(a)}{f(b) - f(a)} \end{aligned}$$

Although you can learn this formula off by heart it is often easier to derive it from first principles using the equation of a straight line or even similar triangles.

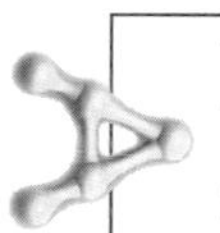

When a root of the equation $f(x) = 0$ is known to lie between $x = a$ and $x = b$, linear interpolation involves replacing the curve by a straight line and gives an approximation to the root as

$$\frac{af(b) - bf(a)}{f(b) - f(a)}.$$

Worked example 10.6

Show that the equation $x^3 - 4x - 10 = 0$ has root between 2 and 3.

Use linear interpolation to obtain an approximation to this root.

Solution

Writing $f(x) = x^3 - 4x - 10$ gives $f(2) = -10$ and $f(3) = 5$.

Using the linear interpolation formula gives the approximation

$$\frac{(5 \times 2) - (-10 \times 3)}{5 - (-10)} = \frac{40}{15} \approx 2.7$$

Alternatively, the straight line through $(2, -10)$ and $(3, 5)$ has gradient 15 and hence has equation $y = 15x - 40$.

Therefore, when $y = 0$, $x = \dfrac{40}{15} = \dfrac{8}{3} \approx 2.7$.

A simple diagram and using similar triangles is just as effective.

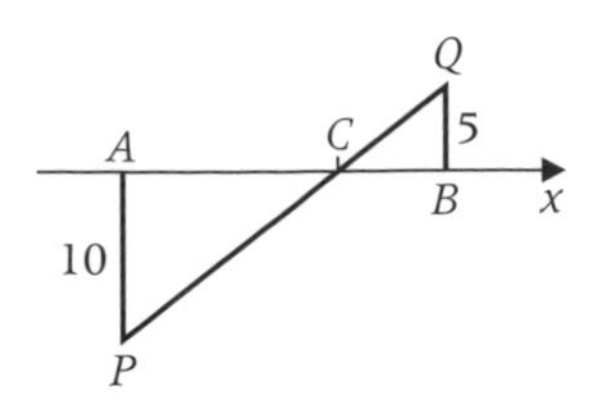

Since $AC:CB = 10:5 = 2:1$, the point C is $\frac{2}{3}$ the way along AB. But A is $(2, 0)$ and B is $(3, 0)$, so the x-coordinate of C is $2 + \frac{2}{3} = 2\frac{2}{3}$.

(The actual value of the root is 2.7608... and this value could easily be obtained by repeating the linear interpolation process.)

EXERCISE 10C

1 (a) Show that the equation $x^3 + 2x - 1 = 0$ has a root between 0 and 1.

(b) Find an approximation to the root by using linear interpolation between 0 and 1.

2 Show that the equation $x^5 + 4x - 50 = 0$ has a root between 2 and 3.
Use linear interpolation to find an approximation to this root.

3 Show that the equation $2^x - 2x - 5 = 0$ has a root between 3 and 4.
Use linear interpolation to find an approximation to this root.

4 Show that the equation $x^3 + 4x^2 + 1 = 0$ has a root between -4.1 and -3.9.
Use linear interpolation to find an approximation to this root.

5 Show that the equation $\sin x + 3x - 5 = 0$ has a root between 1 and 2.
Use linear interpolation to find an approximation to this root.

6 Show that the equation $\cos x - 2x + 3 = 0$ has a root between 1 and 2.
Use linear interpolation to find an approximation to this root.

10.4 The Newton–Raphson iterative formula

The previously mentioned techniques for finding roots of equations are quite slow and fairly tedious.

An iterative formula that uses calculus techniques was produced by Isaac Newton and modified by Raphson and is known as the Newton–Raphson method.

It is used to find approximate solutions to the equation $f(x) = 0$. The method is based on the following diagram, making use of the tangent drawn to the curve with equation $y = f(x)$.

The point P lies on the curve with equation $y = f(x)$ and the tangent to the curve at P cuts the x-axis at R. Then, in general, if the point Q is a first approximation to the point S, where the curve crosses the x-axis, a better approximation is that the curve crosses the x-axis at R.

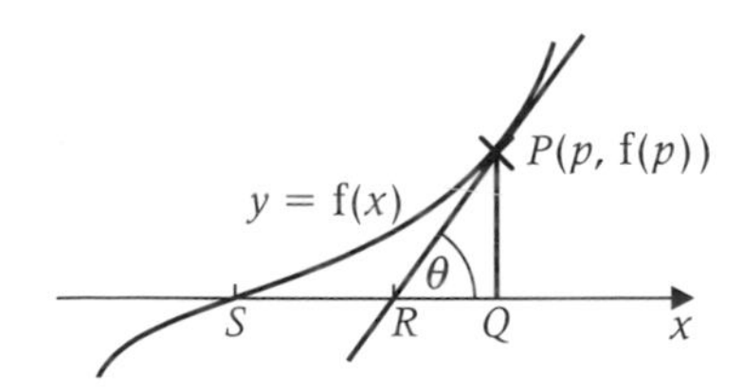

Let the x-coordinate of P be p.

So the y-coordinate of P is $f(p)$. Hence $PQ = f(p)$.

The gradient of the curve at P is given by $f'(p)$.

But the gradient of the tangent is also $\tan\theta$, and from the diagram $\tan\theta = \frac{PQ}{QR}$. Hence $f'(p) = \frac{PQ}{QR} = \frac{f(p)}{QR}$.

Rearranging gives $QR = \frac{f(p)}{f'(p)}$.

So, if the x-coordinate of Q is p, then the x-coordinate of R is

$$p - \frac{f(p)}{f'(p)}.$$

This is, in effect, the Newton–Raphson formula, namely that if p is an approximation to a root of $f(x) = 0$, then usually a better approximation to the root is $p - \frac{f(p)}{f'(p)}$.

This can be expressed as an iterative formula, where successive approximations are x_n and x_{n+1}.

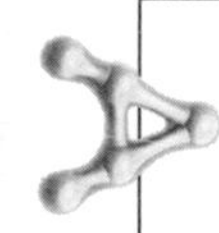

The Newton–Raphson iterative formula for solving $f(x) = 0$ is $x_{n+1} = x_n - \frac{f(x_n)}{f'(x_n)}$.

This formula is given in the booklet for use in the examination under the heading *Numerical Solution of Equations.*

Worked example 10.7

Show, by sketching two graphs on the same axes that the equation $x^3 + x - 1 = 0$ has a single root.

Use the Newton–Raphson method, with first approximation $x_1 = 0.7$, to find further approximations x_2 and x_3, giving your answers to four decimal places. Show on a diagram how convergence to the root takes place.

Solution

Drawing the graphs of $y = x^3 - 1$ and $y = -x$ on the same axes, they are seen to cross exactly once and so the equation $x^3 + x - 1 = 0$ has a single root.

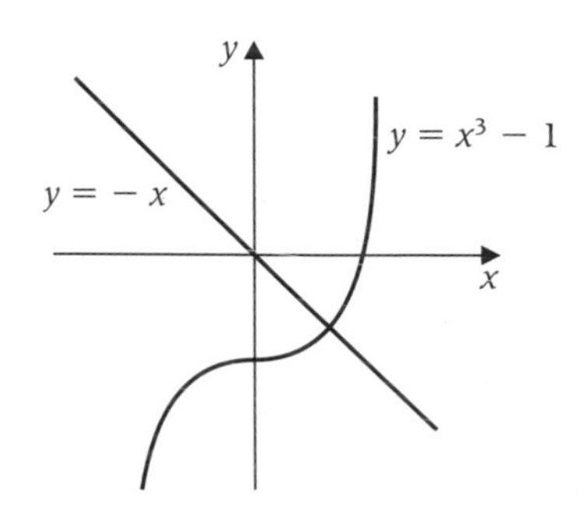

Now let $f(x) = x^3 + x - 1$.

Differentiating gives $f'(x) = 3x^2 + 1$.

Since the Newton–Raphson formula is $x_{n+1} = x_n - \frac{f(x_n)}{f'(x_n)}$

and $x_1 = 0.7$, you have $x_2 = x_1 - \frac{f(x_1)}{f'(x_1)} = 0.7 - \frac{f(0.7)}{f'(0.7)}$.

$f(0.7) = (0.7)^3 + 0.7 - 1 = 0.043$
$f'(0.7) = 3(0.7)^2 + 1 = 2.47$
Hence $x_2 = 0.682\,591\ldots$.

In order to find x_3, you need to use

$$x_3 = x_2 - \frac{f(x_2)}{f'(x_2)} = 0.682\,591 - \frac{f(0.682\,591)}{f'(0.682\,591)}$$

which gives $0.682\,327\,8\ldots$.

Hence $x_2 = 0.6826$ and $x_3 = 0.6823$ (4 d.p.).

You could use the ANS key on your graphics calculator effectively.
Input 0.7 then EXE followed by ANS − (ANS^3 + ANS − 1) ÷ (3*ANS^2 + 1), then EXE for as many times as you need iterations.

The question asked for the answers to be given to four decimal places. It is always better to work with more figures and then to finally round your answers to the required accuracy.

The iterations take place as shown in the diagram. The tangent is drawn to the curve where $x = x_1$ and this tangent meets the x-axis at $x = x_2$, where another tangent is drawn to the curve and the process repeated.

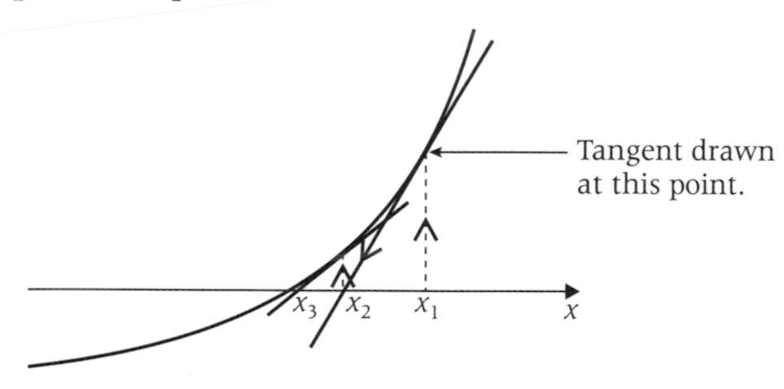

EXERCISE 10D

1 Use the Newton–Raphson method with the value of x_1 given in order to find the next approximation, x_2, to a root of the given equations, giving your answers to three significant figures:

(a) $x^3 - 5x + 11 = 0, \quad x_1 = -3$,

(b) $x^4 - 5x^3 + 6 = 0, \quad x_1 = 1$,

(c) $x^5 - 2x^3 - 17 = 0, \quad x_1 = 2$.

2 Two students are attempting to use Newton–Raphson's method to solve the equation $x^3 - 3x + 4 = 0$. Student A decides to use $x_1 = -1$ and student B uses $x_1 = -3$ as a first approximation. Explain why one of the students will be successful in finding the value of the root and the other will not.

3 The equation $4x^3 - 5x^2 + 2 = 0$ has a single root α. Use the Newton–Raphson iterative method with a first approximation $x_1 = -0.5$ to find the next two successive approximations for α, giving your answers to five decimal places.

4 (a) Prove that the equation $x^3 - 5x + 7 = 0$ has a root α between -3 and -2.

(b) The Newton–Raphson method is to be used to find an approximation for α. Use $x_1 = -3$ as a first approximation to find the value of x_2 giving your answer correct to three decimal places. [A]

5 A curve has equation $y = x^2 + 5 + \frac{2}{x}$ and it crosses the x-axis at the single point $(\alpha, 0)$.

(a) Show that α lies between -0.4 and -0.3.

(b) Use linear interpolation once to find a further approximation, giving your answer to two decimal places.

(c) Use Newton–Raphson's iterative method with first approximation $x_1 = -0.4$ to find the values of x_2 and x_3, giving your answers to five decimal places.

6 The equation $x^5 - 5x + 6 = 0$ has a single root α.

(a) Show that α lies between -1 and -2.

(b) Use the bisection method to find an interval of width 0.25 in which the root lies.

(c) Use linear interpolation once to find an approximation to the value of α, giving your answer to one decimal place.

(d) Use Newton–Raphson's iterative method with first approximation $x_1 = -2$ to find further approximations x_2 and x_3, giving your answers to two decimal places.

10.5 Numerical method to find a point on a curve

You learnt in C1 how to find the equation of a curve passing through a given point where the derivative was known.

For instance to find the curve passing through (2, 5) which satisfies the differential equation $\frac{dy}{dx} = 3x^2 + 1$, you simply integrate to obtain $y = x^3 + x + c$, where c is a constant. Since the curve passes through the point (2, 5), then $5 = 8 + 2 + c$.

Therefore $c = -5$ and so the curve has equation $y = x^3 + x - 5$.

You could then use this equation to find the value of y when $x = 2.1$, 2.2 or for any other value of x you choose.

Suppose, however the curve satisfies the differential equation $\frac{dy}{dx} = \sqrt{(3x^2 + 1)}$. You may struggle at this stage in your course to integrate the expression $\sqrt{(3x^2 + 1)}$. However, a numerical method could be used to estimate the value of y, when $x = 2.1$, if you know that this curve passes through the point (2, 5).

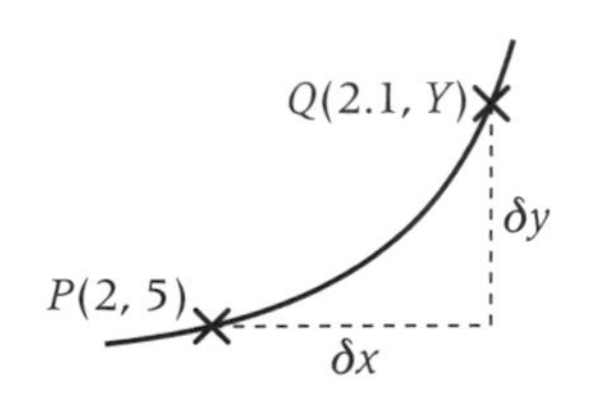

The gradient of the curve at P is given by the value of $\frac{dy}{dx}$ when $x = 2$.

But if the point Q is fairly close to P, then $\frac{\delta y}{\delta x} \approx \frac{dy}{dx}$ and hence $\delta y \approx \frac{dy}{dx} \times \delta x$.

For this curve, $\frac{dy}{dx} = \sqrt{(3x^2 + 1)}$.

At P, $\frac{dy}{dx} = \sqrt{(3 \times 4 + 1)} = \sqrt{13} \approx 3.605\,55$.

The change in x-coordinate is 0.1, so $\delta x = 0.1$

Hence $\delta y \approx \frac{dy}{dx} \times \delta x \approx 3.605\,55 \times 0.1 \approx 0.361$ (3 s.f.).

The y-coordinate of Q is, therefore, given by $Y = 5 + \delta y$.

Hence $Y \approx 5.361$.

10.6 Euler's step-by-step method

The technique introduced in the previous section can be continued for as many steps as you like. The method is attributed to Euler and can be formulated as below.

The step size, δx, has the same value for each step and is denoted by h, which is assumed to be small.

The successive values of x are denoted by x_n and x_{n+1}, and the corresponding values of y by y_n and y_{n+1}.

The initial point through which the curve passes is usually denoted by (x_0, y_0).

Euler's method is used to find a numerical solution of the differential equation $\frac{dy}{dx} = f(x)$. The formula to find y is given by $y_{n+1} = y_n + hf(x_n)$, where $x_{n+1} = x_n + h$.

This formula is in the booklet for use in the examination under the heading 'Numerical solution of differential equations'.

Worked example 10.8

A curve satisfies the differential equation $\frac{dy}{dx} = \frac{4}{x+1}$. Starting at the point (3, 5) on the curve, use a step-by-step method with a step length of 0.2 to estimate the value of y at $x = 3.4$, giving your answer to two decimal places.

Discuss how you could obtain a more accurate estimate for the value of y at $x = 3.4$.

Solution

From the information given, $h = 0.2$, $x_0 = 3$, $y_0 = 5$ and $f(x) = \frac{4}{x+1}$.

Using $x_{n+1} = x_n + h$, then $x_1 = 3.2$ and $x_2 = 3.4$.

Since $y_{n+1} = y_n + hf(x_n)$, you can find

$$y_1 = y_0 + 0.2 \times f(x_0) = 5 + 0.2 \times \frac{4}{x_0 + 1} = 5 + \frac{0.8}{4} = 5.2$$

$$y_2 = y_1 + 0.2 \times f(x_1) = 5.2 + 0.2 \times \frac{4}{x_1 + 1} = 5.2 + \frac{0.8}{4.2} = 5.39 \text{ (2 d.p.)}$$

The accuracy could be improved by increasing the number of steps (or decreasing the step size).

Suppose you reduce the step size to 0.1, you could set out the working in a table for convenience.

$f(x_n) = \frac{4}{x_n + 1}$

n	x_n	y_n	$f(x_n)$	h	$hf(x_n)$
0	3	5	1	0.1	0.1
1	3.1	5.1	0.9756	0.1	0.097 56
2	3.2	5.197 56	0.9524	0.1	0.095 24
3	3.3	5.292 80	0.9302	0.1	0.093 02
4	3.4	5.385 82			

The value of y_2 is given by $y_2 = y_1 + 0.1 \times f(x_1)$.

The estimate for y when $x = 3.4$, using a step size of 0.1 is 5.386 (3 d.p.).

The exact value is known to be 5.381 24….

By reducing the step size even more you would soon get a very accurate estimate. Why not try it on a spreadsheet?

Worked examination question 10.9

A curve satisfies the differential equation $\frac{dy}{dx} = \frac{1}{x^3 + 1}$. Starting at the point (1, 0.5) on the curve, use a step-by-step method with a step length of 0.25 to estimate the value of y at $x = 1.5$, giving your answer to two decimal places.

Solution

From the information given, $h = 0.25$, $x_0 = 1$, $y_0 = 0.5$ and $f(x) = \frac{1}{x^3 + 1}$.

Using $x_{n+1} = x_n + h$, then $x_1 = 1.25$ and $x_2 = 1.5$.

Since $y_{n+1} = y_n + hf(x_n)$, you can find

$$y_1 = y_0 + 0.25 \times f(x_0) = 0.5 + 0.25 \times \frac{1}{1+1} = 0.5 + 0.125 = 0.625$$

$$y_2 = y_1 + 0.25 \times f(x_1) = 0.625 + 0.25 \times \frac{1}{1.25^3 + 1} = 0.7096\ldots$$

The value of y is approximately 0.71 (2 d.p.)

10

It is quite in order to use a tabular form in an exam:

n	x_n	y_n	$f(x_n)$	h	$hf(x_n)$
0	1	0.5	0.5	0.25	0.125
1	1.25	0.625	0.3386	0.25	0.084 66
2	1.5	0.709 66			

which you may prefer to simplify to:

x	y	$\frac{dy}{dx}$	δx	δy
1	0.5	0.5	0.25	0.125
1.25	0.625	0.3386	0.25	0.084 66
1.5	0.709 66			

Whichever tabular form you prefer, you must give a final statement. In this case:

The estimate of y when $x = 1.5$ is 0.71 (to two decimal places).

EXERCISE 10E

1 A curve satisfies the differential equation $\frac{dy}{dx} = \frac{1}{x^2 + 3}$.
Starting at the point (1, 2) on the curve, use a step-by-step method with a step length of 0.25 to estimate the value of y at $x = 1.5$, giving your answer to two decimal places.

2 A curve satisfies the differential equation $\frac{dy}{dx} = \sqrt{x^3 + 1}$.
Starting at the point (2, 4) on the curve, use a step-by-step method with a step length of 0.1 to estimate the value of y at $x = 2.3$, giving your answer to three decimal places.

3 A curve satisfies the differential equation $\frac{dy}{dx} = \frac{1}{3 - x}$.
Starting at the point (1, 4) on the curve, use a step-by-step method with a step length of 0.2 to estimate the value of y at $x = 1.6$, giving your answer to three decimal places.

4 A curve satisfies the differential equation $\frac{dy}{dx} = \sqrt{x^4 + 9}$.
Starting at the point (2, −1) on the curve, use a step-by-step method with a step length of 0.1 to estimate the value of y at $x = 2.5$, giving your answer to three decimal places.

5 A curve satisfies the differential equation $\frac{dy}{dx} = \frac{x}{x + 2}$.
Starting at the point (3, 0.7) on the curve, use a step-by-step method with a step length of 0.125 to estimate the value of y at $x = 3.5$, giving your answer to three decimal places.

6 A curve satisfies the differential equation $\frac{dy}{dx} = \frac{x}{\sqrt{x^2 + 5}}$.

(a) Starting at the point (2, 0) on the curve, use a step-by-step method with a step length of 0.2 to estimate the value of y at $x = 2.4$, giving your answer to two decimal places.

(b) Starting at the point (2, 0) on the curve, use a step-by-step method with a step length of 0.1 to estimate the value of y at $x = 2.4$, giving your answer to three decimal places.

(c) Given that the exact value of y at $x = 2.4$ is 0.280 24 (to five decimal places), determine the percentage errors in your answers to **(a)** and **(b)**.

Key point summary

1 If the graph of $y = f(x)$ is continuous over the interval $a \leqslant x \leqslant b$, and $f(a)$ and $f(b)$ have different signs, then a root of the equation $f(x) = 0$ must lie in the interval $a < x < b$. *p130*

2 When a root of $f(x) = 0$ is known to lie between $x = a$ and $x = b$, the bisection method requires you to next find $f\left(\frac{a+b}{2}\right)$. There must then be a change of sign which allows you to bisect the interval in which the root lies.
The procedure is repeated until you have an interval of the desired width containing the root. *p133*

3 When a root of the equation $f(x) = 0$ is known to lie between $x = a$ and $x = b$, linear interpolation involves replacing the curve by a straight line and gives an approximation to the root as

$$\frac{af(b) - bf(a)}{f(b) - f(a)}.$$

p135

4 The Newton–Raphson iterative formula for solving $f(x) = 0$ is $x_{n+1} = x_n - \frac{f(x_n)}{f'(x_n)}$. *p137*

5 Euler's method is used to find a numerical solution of the differential equation $\frac{dy}{dx} = f(x)$. The formula to find y is given by $y_{n+1} = y_n + hf(x_n)$, where $x_{n+1} = x_n + h$. *p140*

Test yourself	What to review
1 (a) Prove that the equation $x^3 + 10x = 30$ has a root between 2 and 2.1. **(b)** Use the bisection method to find an interval of width 0.025 in which the root lies.	*Section 10.2*
2 Show that the equation $\sin x + 2x - 3 = 0$ has a root between 1.06 and 1.07. Use linear interpolation to find a three decimal place approximation to this root.	*Section 10.3*
3 The equation $x^5 + x - 35 = 0$ has a root close to 2. Use the Newton–Raphson method with $x_1 = 2$ to find the values of x_2 and x_3, giving your answers to five significant figures.	*Section 10.4*
4 A curve satisfies the differential equation $\frac{dy}{dx} = \sqrt{x^2 + 5}$. Starting at the point (2, 6) on the curve, use a step-by-step method with a step length of 0.1 to estimate the value of y at $x = 2.2$, giving your answer to three decimal places.	*Section 10.6*

Test yourself ANSWERS

1 (b) 2.075 and 2.100.

2 1.063.

3 2.0123, 2.0122.

4 6.607.

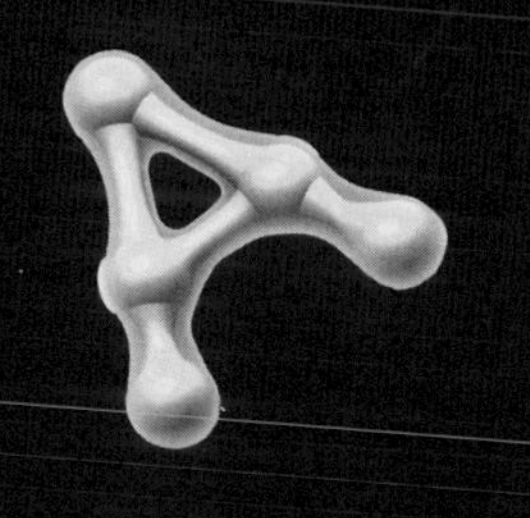

CHAPTER 11

Asymptotes and rational functions of the form $\dfrac{ax + b}{cx + d}$

Learning objectives

After studying this chapter, you should be able to:

- understand what is meant by an asymptote
- find equations of vertical asymptotes
- find equations of horizontal asymptotes for graphs with equations of the form $y = \dfrac{ax + b}{cx + d}$
- sketch curves with equations of the form $y = \dfrac{ax + b}{cx + d}$
- use graphs to solve inequalities involving rational functions.

11.1 Asymptotes

You will be familiar with the graph of $y = \dfrac{1}{x}$.

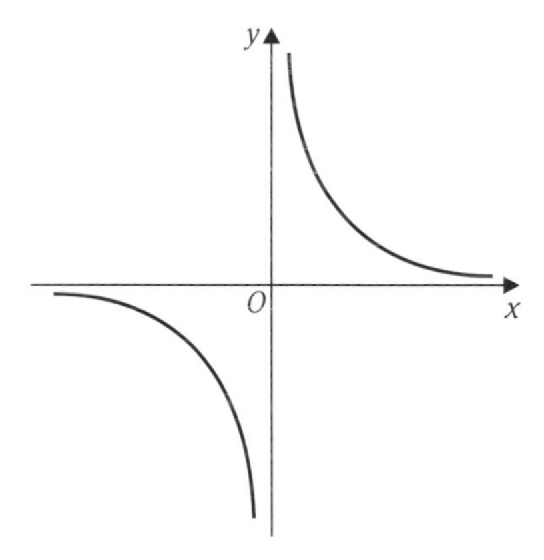

As x becomes larger and larger, we write $x \to \infty$, and we observe that $y \to 0$.

Similarly, as $x \to -\infty$, we see that $y \to 0$.

Because the curve gets closer and closer to the line $y = 0$ without actually touching it, we call $y = 0$ (the x-axis) an asymptote.

The line $x = 0$ (the y-axis) is also an asymptote to this curve since the curve gets closer and closer to this line as $x \to 0$.

The Oxford English Dictionary definition of this unusual word is

asymptote//n.

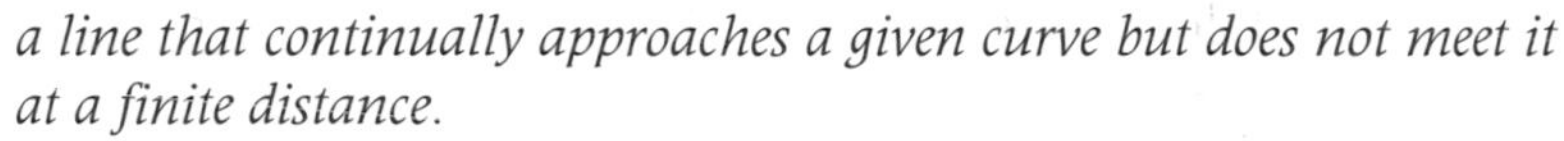

> *a line that continually approaches a given curve but does not meet it at a finite distance.*
> *From Greek 'asumptotos' 'not falling together' (a- 'not' + sum- 'together' + ptotos- 'falling' from pipto- 'fall').*

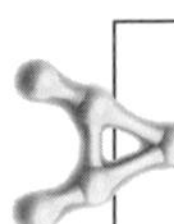

> An asymptote is a line that a curve approaches for large values of $|x|$ or $|y|$. It is usually represented by a broken or dotted line.

When either the x-axis or y-axis is an asymptote, the broken line is not actually drawn over the full line representing the coordinate axis.

A translation of the curve $y = \frac{1}{x}$ through $\begin{bmatrix} 2 \\ 0 \end{bmatrix}$ would give the curve with equation $y = \frac{1}{x - 2}$ and its graph is shown below.

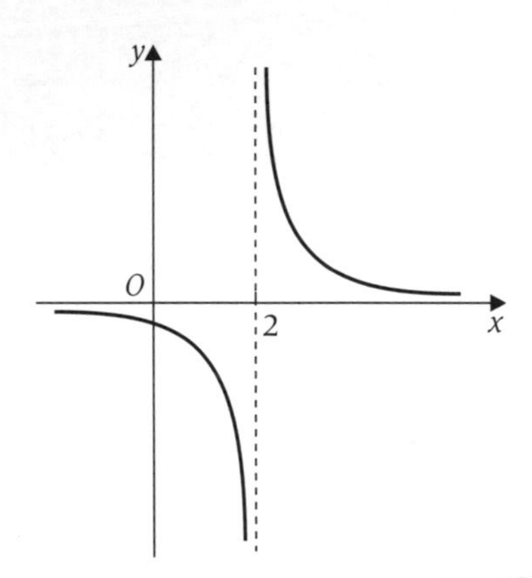

The line $y = 0$ (the x-axis) is still an asymptote. It is referred to as a horizontal asymptote.

The line $x = 2$ is now an asymptote. It is shown as a broken line and is referred to as a vertical asymptote.

When the curve $y = \frac{1}{x}$ is translated through $\begin{bmatrix} 0 \\ 3 \end{bmatrix}$ you obtain the curve with equation $y = \frac{1}{x} + 3$ and its graph is shown below. The horizontal asymptote has equation $y = 3$ and is shown as a broken line. The line $x = 0$ (the y-axis) is a vertical asymptote.

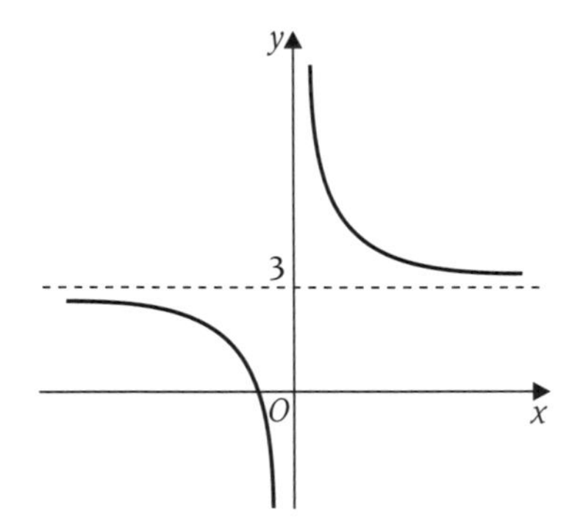

11.2 Vertical asymptotes

When a curve has an equation of the form $y = \frac{f(x)}{g(x)}$, the curve is not defined for any value of x for which $g(x) = 0$.

Whenever there is a value a such that $g(a) = 0$, we say that $x = a$ is a vertical asymptote.

Worked example 11.1

Find the equations of any vertical asymptotes, if they exist, of the following curves:

(a) $y = \dfrac{x^2 + 7}{x^2 - x - 6}$ **(b)** $y = \dfrac{x^2 + x - 7}{x^2 + 3x + 5}$

Solution

(a) The denominator of the rational function is $x^2 - x - 6$.
To find vertical asymptotes you need to solve the equation

$$x^2 - x - 6 = 0.$$

$$\Rightarrow \quad (x - 3)(x + 2) = 0$$

$$\Rightarrow \quad x = 3 \quad \text{or} \quad x = -2$$

The curve $y = \dfrac{x^2 + 7}{x^2 - x - 6}$ has two vertical asymptotes.

They have equations $x = 3$ and $x = -2$.

(b) This time, the denominator of the rational function is

$$x^2 + 3x + 5.$$

The discriminant of the quadratic is $9 - 4 \times 5 = -11$.

Since this is negative, the equation $x^2 + 3x + 5 = 0$ has no real roots.

The curve $y = \dfrac{x^2 + x - 7}{x^2 + 3x + 5}$ has no vertical asymptotes.

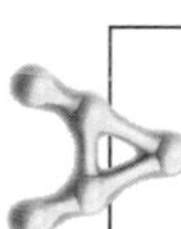

The line $x = a$ is a vertical asymptote of the curve $y = \dfrac{\mathrm{f}(x)}{\mathrm{g}(x)}$ if $\mathrm{g}(a) = 0$.

EXERCISE 11A

Find any vertical asymptotes of the following curves:

1 $y = \dfrac{1}{x - 7}$

2 $y = \dfrac{x + 5}{2x - 4}$

3 $y = \dfrac{x}{x + 1}$

4 $y = \dfrac{1}{x(x - 1)}$

5 $y = \dfrac{x + 5}{(x - 3)(x + 4)}$

6 $y = \dfrac{x}{x^2 + 1}$

7 $y = \dfrac{5}{(x - 1)^2}$

8 $y = \dfrac{3x + 7}{x^2 - 5x - 6}$

9 $y = \dfrac{2x - 4}{x^2 + x + 1}$

11.3 Curves of the form $y = \dfrac{ax+b}{cx+d}$

The curve with equation $y = \dfrac{3x+5}{2x-7}$ has a vertical asymptote.

Its equation is $x = 3\frac{1}{2}$.

The curve also has a horizontal asymptote.

In order to find its equation you should divide throughout by x in both the numerator and denominator.

Clearly when x is very large, the dominant part of $3x + 5$ is $3x$ and the dominant part of $2x - 7$ is $2x$. This suggests that when x is very large in magnitude (either positive or negative), the value of y is approximately $\frac{3x}{2x} = \frac{3}{2}$. It suggests that the horizontal asymptote has equation $y = 1\frac{1}{2}$.

$$y = \frac{3x+5}{2x-7} = \frac{3+\frac{5}{x}}{2-\frac{7}{x}}$$

Now, as $x \to \infty$, $\frac{5}{x} \to 0$ and $\frac{7}{x} \to 0$.

Also, when $x \to -\infty$, $\frac{5}{x} \to 0$ and $\frac{7}{x} \to 0$.

Hence $y \to \frac{3+0}{2-0}$, as $x \to \pm\infty$ (or as $|x| \to \infty$)

and therefore $y = 1\frac{1}{2}$ is an asymptote.

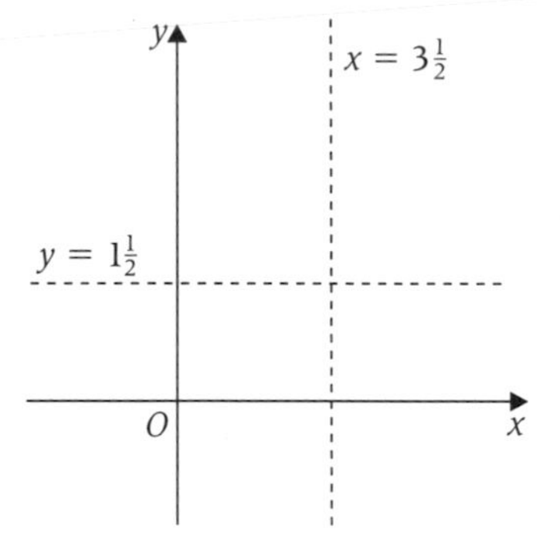

In order to sketch the graph of $y = \dfrac{3x+5}{2x-7}$ you start by inserting the asymptotes.

You can also find when the curve crosses the coordinate axes.

When $x = 0$, $y = \frac{-5}{7}$. This point has coordinates $\left(0, -\frac{5}{7}\right)$.

To find when $y = 0$, put $3x + 5 = 0$. Hence the point $\left(-\frac{5}{3}, 0\right)$ lies on the curve.

You then need to look at values of x just to the right and just to the left of the vertical asymptote.

When $x = 3.4$, $y = \dfrac{3x+5}{2x-7} = \dfrac{10.2+5}{-0.2} = -76$.

The important thing is that it is large and negative.

When $x = 3.6$, $y = \dfrac{3x+5}{2x-7} = \dfrac{10.8+5}{0.2} = 79$.

The important thing is that it is large and positive.

So the behaviour near the vertical asymptote is as shown opposite.

You can now sketch the entire curve.

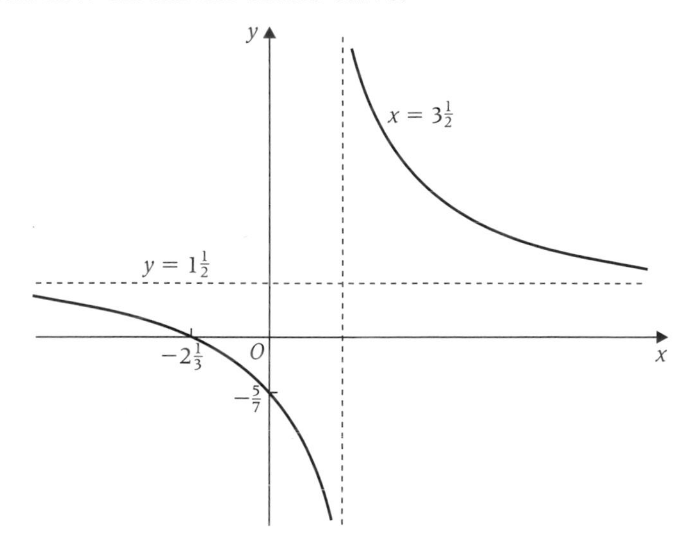

To find a horizontal asymptote of $y = \dfrac{ax+b}{cx+d}$,

re-write the equation as $y = \dfrac{a + \dfrac{b}{x}}{c + \dfrac{d}{x}}$.

As $|x| \to \infty$, $\dfrac{1}{x} \to 0$, therefore $y \to \dfrac{a}{c}$.

The horizontal asymptote has equation $y = \dfrac{a}{c}$.

Note that a common mistake is to write the equation as $y \to \dfrac{a}{c}$.
This is not correct. An equation must have the 'equals' sign.

Worked example 11.2

A curve has equation $y = \dfrac{4-2x}{x+1}$.

(a) State the coordinates of the points where the curve crosses the coordinate axes.

(b) Find the equations of the asymptotes.

(c) Sketch the curve.

Solution

(a) Since $y = \dfrac{4-2x}{x+1}$, $y = 0$ when $4 = 2x$ or when $x = 2$. The curve therefore crosses the x-axis at $(2, 0)$.

When $x = 0$, $y = \dfrac{4-0}{0+1} = 4$.

The curve crosses the y-axis at $(0, 4)$.

(b) The vertical asymptote is found by considering when the denominator of $\frac{4-2x}{x+1}$ is zero.

The vertical asymptote has equation $x = -1$.

For the horizontal asymptote, consider $y = \frac{4-2x}{x+1} = \frac{\frac{4}{x}-2}{1+\frac{1}{x}}$.

Hence $y \to \frac{0-2}{1+0} = -2$, as $x \to \pm\infty$.

The horizontal asymptote has equation $y = -2$.

(c) You can draw in the asymptotes and sketch the curve.

Notice that when $x = -0.9$,
$y = \frac{4-(-1.8)}{-0.9+1} = \frac{5.8}{0.1} = 58.$
Also, when $x = -1.1$,
$y = \frac{4-(-2.2)}{-1.1+1} = \frac{6.2}{-0.1} = -62.$

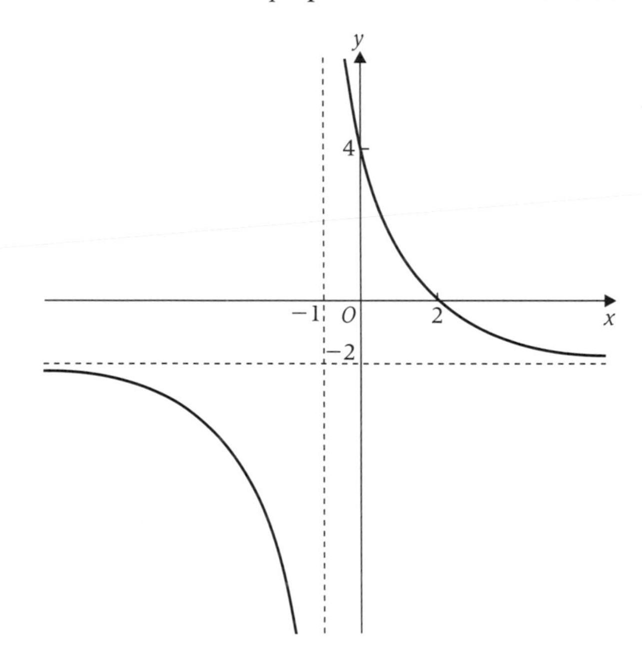

EXERCISE 11B

1 A curve has equation $y = \frac{5-x}{x+2}$.

(a) State the coordinates of the points where the curve crosses the coordinate axes.

(b) Find the equations of the asymptotes.

(c) Sketch the curve.

2 A curve has equation $y = \frac{x+1}{3x-2}$.

(a) State the coordinates of the points where the curve crosses the coordinate axes.

(b) Find the equations of the asymptotes.

(c) Sketch the curve.

3 For each of the following curves:

(i) find the equations of the asymptotes;

(ii) determine the coordinates of the points where the curve cuts the coordinate axes;

(iii) sketch its graph.

(a) $y = \frac{x+3}{2x-8}$, **(b)** $y = \frac{6x-12}{1+2x}$, **(c)** $y = 1 + \frac{3}{x-2}$.

4 For each of the following curves:

(i) find the equations of the asymptotes;

(ii) determine the coordinates of the points where the curve cuts the coordinate axes;

(iii) sketch its graph.

(a) $y = \frac{x-3}{2x+5}$, **(b)** $y = \frac{3x+4}{1-3x}$, **(c)** $y = 3 - \frac{2}{x+2}$.

5 Sketch the graphs of:

(a) $y = \frac{5x+3}{4x-7}$, **(b)** $y = \frac{4-8x}{3-5x}$, **(c)** $y = 2 - \frac{5x+3}{3x-2}$.

State the equations of any asymptotes and points where the curve crosses the coordinate axes.

6 The graph of $y = \mathrm{f}(x)$ is sketched below.
The asymptotes have equations $x = 1$ and $y = 4$.

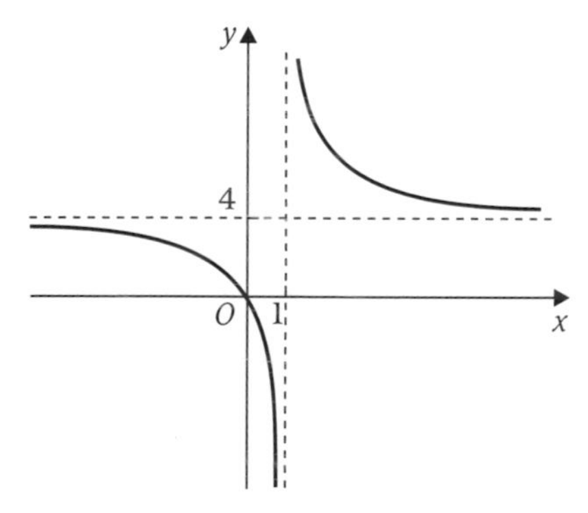

Given that $\mathrm{f}(x) = \frac{ax}{x-b}$, use the sketch to find the values of a and b. [A]

11.4 Finding points of intersection of graphs of rational functions and straight lines

You have already learnt how to find the points where a curve such as $y = \frac{4-2x}{x+1}$ cuts the coordinate axes.

You may need to know where it crosses other straight lines.

Worked example 11.3

Determine the points of intersection of the curve $y = \frac{4 - 2x}{x + 1}$ and the line $y = 3x - 2$.

Solution

At a point of intersection, $\frac{4 - 2x}{x + 1} = 3x - 2$.

Multiplying both sides by $x + 1$ gives $4 - 2x = (3x - 2)(x + 1)$.

Hence $4 - 2x = 3x^2 + x - 2 \Rightarrow 3x^2 + 3x - 6 = 0$

$\Rightarrow x^2 + x - 2 = 0 \Rightarrow (x + 2)(x - 1) = 0$

Therefore $x = 1$ or $x = -2$.

When $x = 1$, $y = 1$.

When $x = -2$, $y = -8$.

You can substitute the value of x into either the equation of the curve or the equation of the line.

The points of intersection are $(1, 1)$ and $(-2, -8)$.

EXERCISE 11C

Find the points of intersection of the following curves and lines:

1 $y = \frac{2}{x + 1}$, $y = 2x - 1$.

2 $y = \frac{x - 1}{2x + 3}$, $y = 3x + 1$.

3 $y = \frac{5x}{2x + 1}$, $y = 4 - x$.

4 $y = \frac{6x}{2x + 5}$, $y = 3x + 1$.

5 $y = \frac{3x - 1}{5x - 3}$, $y = 4x - 3$.

11.5 The use of graphs to solve inequalities

Once you have drawn the graph of a rational function you can use this graph to solve an inequality.

Worked example 11.4

(a) Sketch the graph of $y = \frac{4 - 2x}{x + 1}$.

(b) Hence solve the inequality $\frac{4 - 2x}{x + 1} \leqslant 3$.

Solution

The graph was sketched in Worked example 11.2 and is reproduced below with the line $y = 3$.

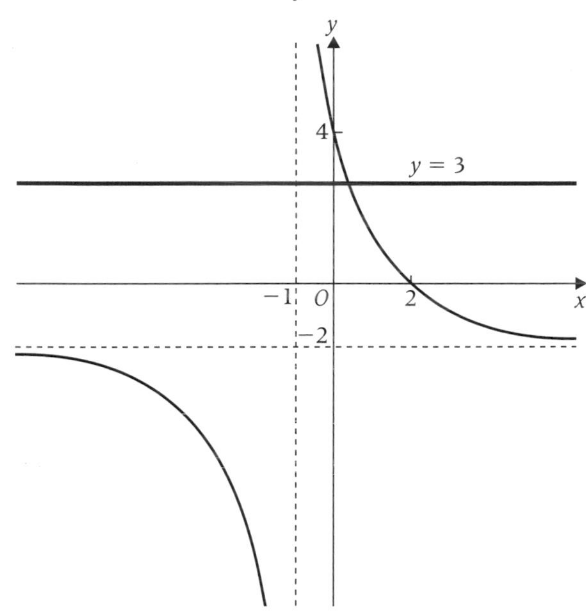

Consider when the curve has a y-value that is less than or equal to 3.

Solving $\dfrac{4-2x}{x+1} = 3$ gives $4 - 2x = 3(x + 1) = 3x + 3$.

$4 - 3 = 3x + 2x \quad \Rightarrow \quad 1 = 5x \quad \Rightarrow \quad x = \frac{1}{5}$

Clearly, from the graph, part of the solution is that $x \geqslant \frac{1}{5}$.

However, another region when the curve lies below the line is $x < -1$, so the complete solution is $x \geqslant \frac{1}{5}$, $x < -1$.

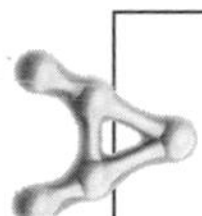

In order to solve inequalities such as

$$\frac{ax+b}{cx+d} < k \quad \text{or} \quad \frac{ax+b}{cx+d} > k:$$

1 Sketch the graph of $y = \dfrac{ax+b}{cx+d}$ and the line $y = k$.

2 Solve the equation $\dfrac{ax+b}{cx+d} = k$.

3 Use the graph to find the possible values of x for which the graph lies below or above the line.

Worked example 11.5

(a) Sketch the graph of $y = \dfrac{3x-1}{x-2}$.

State the coordinates of the points where the curve crosses the coordinate axes and write down the equations of its asymptotes.

(b) Hence, or otherwise, solve the inequality $\dfrac{3x-1}{x-2} > -1$.

11

Solution

You need to find the equations of the asymptotes and the points where the curve crosses the axes before attempting to sketch the curve.

When $y = 0$, $\frac{3x - 1}{x - 2} = 0 \quad \Rightarrow \quad x = \frac{1}{3}$

A common mistake when finding points where the curve crosses the x-axis is to say that another value of x is when the denominator is zero, namely $x = 2$.

When $x = 0$, $y = \frac{0 - 1}{0 - 2} = \frac{1}{2}$.

The coordinates are therefore $\left(\frac{1}{3}, 0\right)$ and $\left(0, \frac{1}{2}\right)$.

The vertical asymptote has equation $x = 2$.

Since $y = \frac{3x - 1}{x - 2} = \frac{3 - \frac{1}{x}}{1 - \frac{2}{x}}$, $y \to 3$ as $|x| \to \infty$.

The horizontal asymptote has equation $y = 3$.

The graph is sketched below:

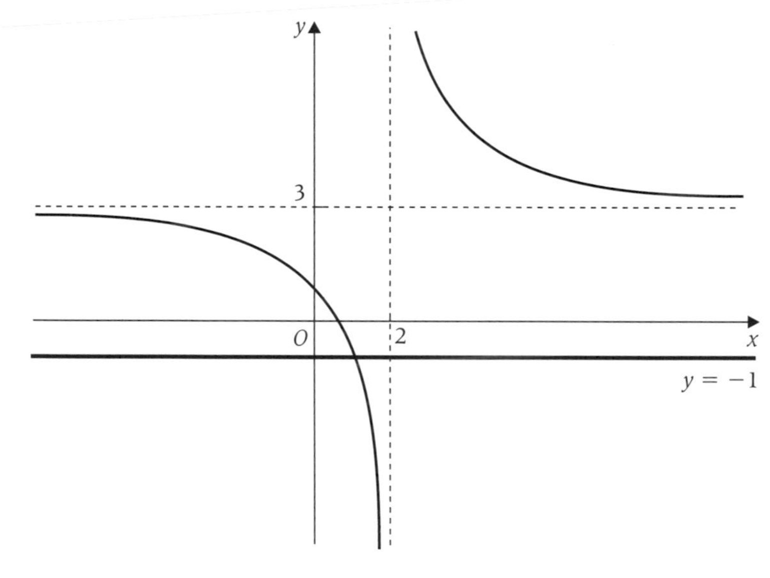

The line $y = -1$ is also drawn and the curve intersects the line when $\frac{3x - 1}{x - 2} = -1 \quad \Rightarrow \quad 3x - 1 = -x + 2 \quad \Rightarrow \quad 4x = 3$.

Hence $x = \frac{3}{4}$.

In order to solve $\frac{3x - 1}{x - 2} > -1$, you need to consider the regions where the curve is above the line.

There are two zones. Hence the solution to the inequality is

$$x < \frac{3}{4}, \quad x > 2.$$

Note that you can check your answer by testing some values you believe are part of the solution, for example:

when $x = 3$, $\frac{3x-1}{x-2} = \frac{9-1}{3-2} = 8 > -1$;

when $x = 0$, $\frac{3x-1}{x-2} = \frac{0-1}{0-2} = \frac{1}{2} > -1$.

EXERCISE 11D

Use the graphs from your answers to exercise 11B to solve the inequalities in questions **1** to **8**:

1 $\frac{5-x}{x+2} < 6$

2 $\frac{x+1}{3x-2} \geqslant \frac{1}{2}$

3 $\frac{x+3}{2x-8} > -3$

4 $\frac{6x-12}{1+2x} > 0$

5 $\frac{x-3}{2x+5} < -5$

6 $\frac{3x+4}{1-3x} > 1$

7 $\frac{5x+3}{4x-7} \geqslant 2$

8 $\frac{4-8x}{3-5x} > 1$

9 **(a)** Sketch the graph of $y = \frac{3-4x}{2x-5}$, stating the coordinates of the points where the curve crosses the coordinate axes and writing down the equations of its asymptotes.

(b) Hence, or otherwise, solve the inequality $\frac{3-4x}{2x-5} < 0$. [A]

10 **(a)** Sketch the graph of $y = \frac{3x+4}{x-2}$, stating the coordinates of the points where the curve crosses the coordinate axes and writing down the equations of its asymptotes.

(b) Hence, or otherwise, solve the inequality $\frac{3x+4}{x-2} > 1$. [A]

11 **(a)** Sketch the curve with equation $y = \frac{4x-3}{x-1}$, stating the coordinates of the points where the curve crosses the coordinate axes and writing down the equations of its asymptotes.

(b) Solve the inequality $\frac{4x-3}{x-1} < x+3$. [A]

Key point summary

1 An asymptote is a line that a curve approaches for large values of $|x|$ or $|y|$. It is usually represented by a broken or dotted line. *p145*

2 The line $x = a$ is a vertical asymptote of the curve $y = \frac{f(x)}{g(x)}$ if $g(a) = 0$. *p147*

3 To find a horizontal asymptote of $y = \frac{ax+b}{cx+d}$, re-write the equation as $y = \frac{a + \frac{b}{x}}{c + \frac{d}{x}}$. *p149*

As $|x| \to \infty$, $\frac{1}{x} \to 0$, therefore $y \to \frac{a}{c}$.

The horizontal asymptote has equation $y = \frac{a}{c}$.

4 In order to solve inequalities such as *p153*

$$\frac{ax+b}{cx+d} < k \quad \text{or} \quad \frac{ax+b}{cx+d} > k:$$

1. Sketch the graph of $y = \frac{ax+b}{cx+d}$ and the line $y = k$.
2. Solve the equation $\frac{ax+b}{cx+d} = k$.
3. Use the graph to find the possible values of x for which the graph lies below or above the line.

Test yourself

What to review

1 Find any vertical asymptotes of the following curves: *Section 11.2*

(a) $y = \frac{x+3}{5x-10}$, **(b)** $y = \frac{3x+5}{(x-4)(x+1)}$, **(c)** $y = \frac{3x}{x^2+4}$

2 A curve has equation $y = \frac{2x+4}{x+1}$. *Section 11.3*

(a) State the coordinates of the points where the curve crosses the coordinate axes.

(b) Find the equations of the asymptotes.

(c) Sketch the curve.

3 Find the points of intersection of the curve with equation $y = \frac{2x+4}{x+1}$ and the straight line $y = x + 2$. *Section 11.4*

4 Use your graph from question **2** to solve the inequality $\frac{2x+4}{x+1} \leqslant 3$. *Section 11.5*

Test yourself ANSWERS

1 **(a)** $x = 2$; **(b)** $x = 4, x = -1$; **(c)** no vertical asymptotes.

2 **(a)** $(-2, 0)$ and $(0, 4)$; **(b)** $x = -1, y = 2$;

(c)

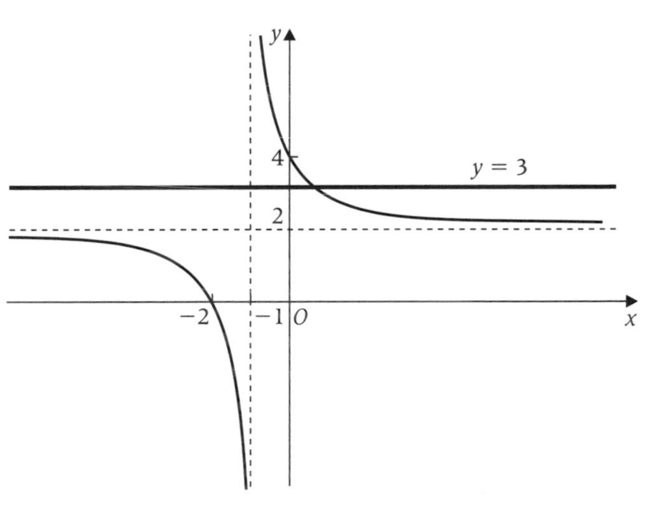

3 $(1, 3)$ and $(-2, 0)$.

4 $x \geqslant 1, \quad x < -1$.

CHAPTER 12

Further rational functions and maximum and minimum points

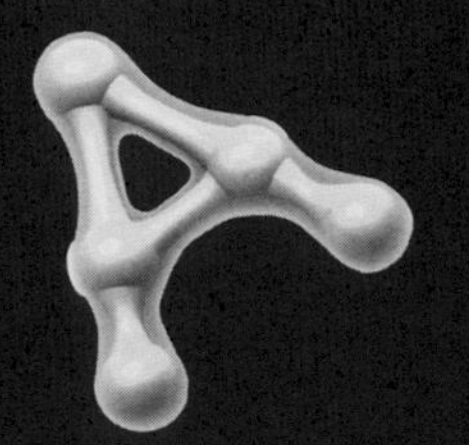

Learning objectives

After studying this chapter, you should be able to:

- find equations of asymptotes
- determine the behaviour close to vertical asymptotes to enable you to sketch the entire curve
- sketch graphs of irrational functions with quadratic denominators
- find regions for which a rational function exists
- use the discriminant to find maximum and minimum points of graphs of certain rational functions.

12.1 Rational functions with quadratic denominators

In the previous chapter you learned to sketch graphs of rational functions with linear denominators.

Suppose you wanted to sketch the graph of $y = \dfrac{1}{(x-2)(x-5)}$.

The vertical asymptotes have equations $x = 2$ and $x = 5$.

As $x \to \infty$, $y \to 0$ and when $x \to -\infty$, $y \to 0$.
The x-axis (the line $y = 0$) is a horizontal asymptote.

You can consider values of x slightly less than 2 and just a little bit more than 2, since $x = 2$ is a vertical asymptote.

When $x = 1.99$, $y = \dfrac{1}{(x-2)(x-5)} \approx \dfrac{1}{-0.01 \times -3} \approx 30$, which is large and positive.

When $x = 2.01$, $y = \dfrac{1}{(x-2)(x-5)} \approx \dfrac{1}{0.01 \times -3} \approx -30$, which is large and negative.

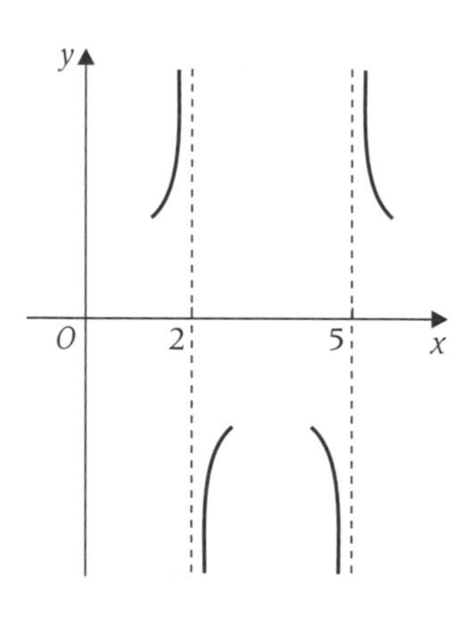

A similar consideration near $x = 5$ leads to the behaviour shown close to each of the vertical asymptotes.

Since the equation $\dfrac{1}{(x-2)(x-5)} = 0$ has no solutions, the curve never crosses the x-axis.

The curve between $x = 2$ and $x = 5$ must reach a maximum point then decrease again.

It does cross the y-axis when $x = 0$. This gives the point $(0, 0.1)$.

You also know that as $|x| \to \infty$, $y \to 0$.

This enables you to sketch the graph of the complete curve.

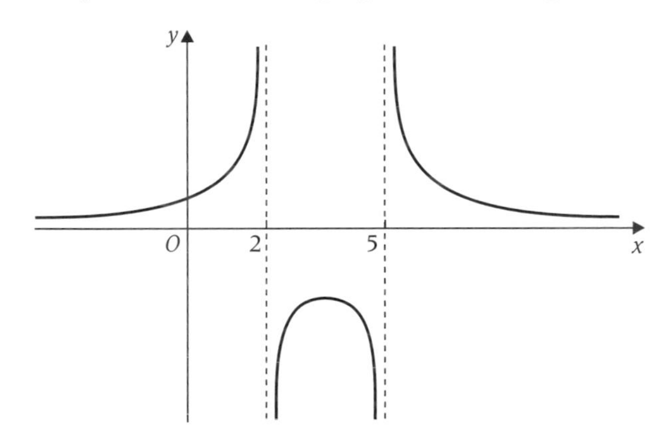

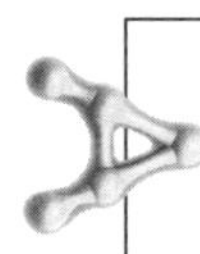

It is always useful to draw any asymptotes as the first stage in sketching the graph of a rational function.

When a curve has a vertical asymptote at $x = a$, it is useful to check the values of y when x is a little smaller than a and when x is a little larger than a.

By considering the behaviour very close to the asymptotes, it is often possible to deduce the main shape of the graph.

12.2 Rational functions of the form $\dfrac{px + q}{ax^2 + bx + c}$

The graph of a function of this form will have two vertical asymptotes whenever $b^2 - 4ac > 0$.

However, when $b^2 - 4ac < 0$, the graph will have no vertical asymptotes.

It will cut the x-axis exactly once at the point where $x = -\dfrac{q}{p}$.

For any function of this type, as $|x| \to \infty$, $y \to 0$.

Therefore $y = 0$ will always be a horizontal asymptote.

Worked example 12.1

A curve has equation $y = \dfrac{12x - 12}{(x + 1)(x - 3)}$.

(a) Write down the equations of its asymptotes.

(b) Sketch the curve.

Solution

(a) The vertical asymptotes are $x = -1$ and $x = 3$. The horizontal asymptote is $y = 0$.

(b) You begin by sketching the asymptotes.

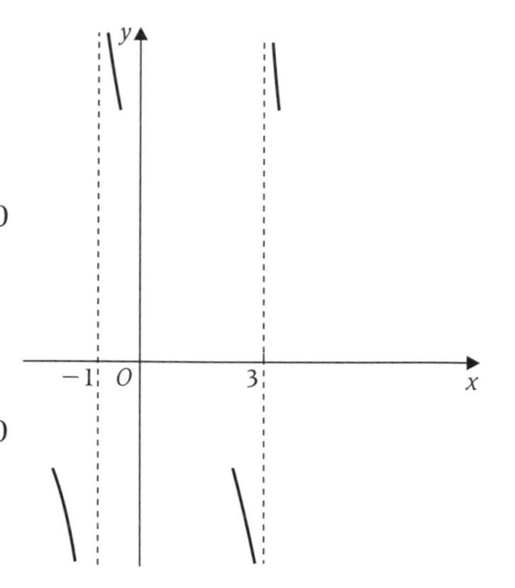

Close to $x = 3$:

When $x = 2.99$, $y = \dfrac{12x - 12}{(x+1)(x-3)} \approx \dfrac{36 - 12}{(3+1) \times -0.01} \approx -600$

When $x = 3.01$, $y = \dfrac{12x - 12}{(x+1)(x-3)} \approx \dfrac{36 - 12}{(3+1) \times 0.01} \approx 600$

Close to $x = -1$

When $x = -0.99$, $y = \dfrac{12x - 12}{(x+1)(x-3)} \approx \dfrac{-12 - 12}{0.01 \times (-1-3)} \approx 600$

When $x = -1.01$, $y \approx \dfrac{-12 - 12}{-0.01 \times (-1-3)} \approx -600$

The curve only crosses the x-axis once: at $(1, 0)$.
It also cuts the y-axis at $(0, 4)$.

You now have enough information to sketch the complete curve.

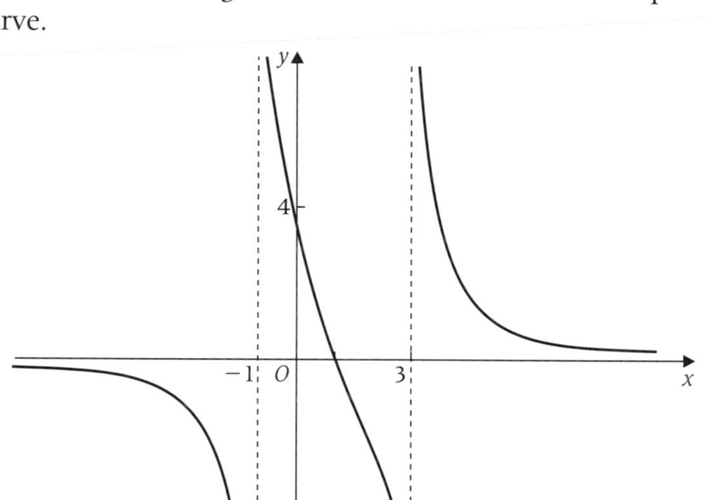

Worked example 12.2

A curve has equation $y = \dfrac{x - 3}{x^2 + x + 3}$.

Write down the equations of its asymptotes.

Solution

The denominator is $x^2 + x + 3$ and the discriminant is negative.
Since $x^2 + x + 3 = 0$ has no real roots, there are no vertical asymptotes.
The only asymptote is the x-axis, or $y = 0$.

It is not easy to sketch the curve until you know the coordinates of any maximum or minimum points.

EXERCISE 12A

For each of the curves, **1** to **6**, find the equations of any asymptotes and sketch the curves.

1 $y = \dfrac{1}{(x-3)(x-4)}$

2 $y = \dfrac{1}{(x-2)(3-x)}$

3 $y = \dfrac{1}{x^2-4}$

4 $y = \dfrac{x+1}{(x-1)(3-x)}$

5 $y = \dfrac{x-5}{x^2-x-6}$

6 $y = \dfrac{2x-3}{x(x-4)}$

7 Find the equations of any asymptotes of the following curves:

(a) $y = \dfrac{x+4}{x^2-3x-10}$, **(b)** $y = \dfrac{x+4}{x^2-3x+10}$.

8 The curve C has equation $y = \dfrac{x+5}{x^2-x+6}$. The line $y = k$ intersects the curve C. Find the points of intersection, if any, in the cases when:

(a) $k = \frac{1}{2}$, **(b)** $k = 1$, **(c)** $k = 2$.

12.3 The use of the discriminant to find regions for which a curve is defined

In the last exercise you found where the curve $y = \dfrac{x+5}{x^2-x+6}$ intersected the line $y = k$ for certain values of k.

In general $\dfrac{x+5}{x^2-x+6} = k$ can be written as $x + 5 = k(x^2 - x + 6)$, which can be re-written as $kx^2 - (k+1)x + (6k-5) = 0$.

By considering the discriminant of the quadratic, you can determine the possible values of k that produce real values of x.

The quadratic equation $ax^2 + bx + c = 0$ has real roots when the discriminant is greater than or equal to zero.
$b^2 - 4ac \geqslant 0$

The discriminant is $(k+1)^2 - 4k(6k-5)$, and for real values of x

$$(k+1)^2 - 4k(6k-5) \geqslant 0.$$

$$k^2 + 2k + 1 - 24k^2 + 20k \geqslant 0$$

$$\Rightarrow \quad -23k^2 + 22k + 1 \geqslant 0$$

$$\Rightarrow \quad 23k^2 - 22k - 1 \leqslant 0$$

$$\Rightarrow \quad (23k+1)(k-1) \leqslant 0$$

The sign diagram for $(23k+1)(k-1)$ is shown opposite.

Hence $-\frac{1}{23} \leqslant k \leqslant 1$ for real values of x.

In order to find the set of values of y for which a curve of the form $y = \dfrac{ax^2 + bx + c}{dx^2 + ex + f}$ exists:

1 consider where $y = k$ cuts the curve by writing $k = \dfrac{ax^2 + bx + c}{dx^2 + ex + f}$;

2 multiply out to obtain a quadratic of the form $Ax^2 + Bx + C = 0$, where A, B and C will involve k;

3 use the condition for real roots $B^2 - 4AC \geqslant 0$ to obtain a quadratic inequality involving k;

4 convert the solution involving k to a condition involving y.

Worked example 12.3

A curve has equation $y = \dfrac{x^2 - 4}{x^2 + 4x + 6}$.

Find the values of y for which the curve exists.

Solution

Consider where the line $y = k$ cuts the curve.

This gives the equation $k = \dfrac{x^2 - 4}{x^2 + 4x + 6}$, which can be multiplied out to give $k(x^2 + 4x + 6) = x^2 - 4$.

Writing as a quadratic in x: $(k - 1)x^2 + 4kx + (6k + 4) = 0$.

For real values of x, the discriminant is greater than or equal to zero.

Hence, $\quad 16k^2 - 4(k - 1)(6k + 4) \geqslant 0$

or $\quad 16k^2 - 4(6k^2 - 2k - 4) \geqslant 0 \quad \Rightarrow \quad -8k^2 + 8k + 16 \geqslant 0$

$\Rightarrow \quad 0 \geqslant 8k^2 - 8k - 16 \quad$ or $\quad k^2 - k - 2 \leqslant 0$

$\Rightarrow \quad (k - 2)(k + 1) \leqslant 0$

The critical values are $k = 2$ and $k = -1$.

You could sketch a graph of $(k - 2)(k + 1)$ but it is quicker to construct a sign diagram

For example, when $k = 0$, $(k - 2)(k + 1) = -2$ which is negative.

	-1		2	
$+$		$-$		$+$

The solution to the inequality is therefore $-1 \leqslant k \leqslant 2$.

The curve $y = \dfrac{x^2 + 4x + 6}{x^2 + 4}$ exists for $-1 \leqslant y \leqslant 2$.

If you were to use your graphics calculator, you would see that the curve looks like the curve shown below.

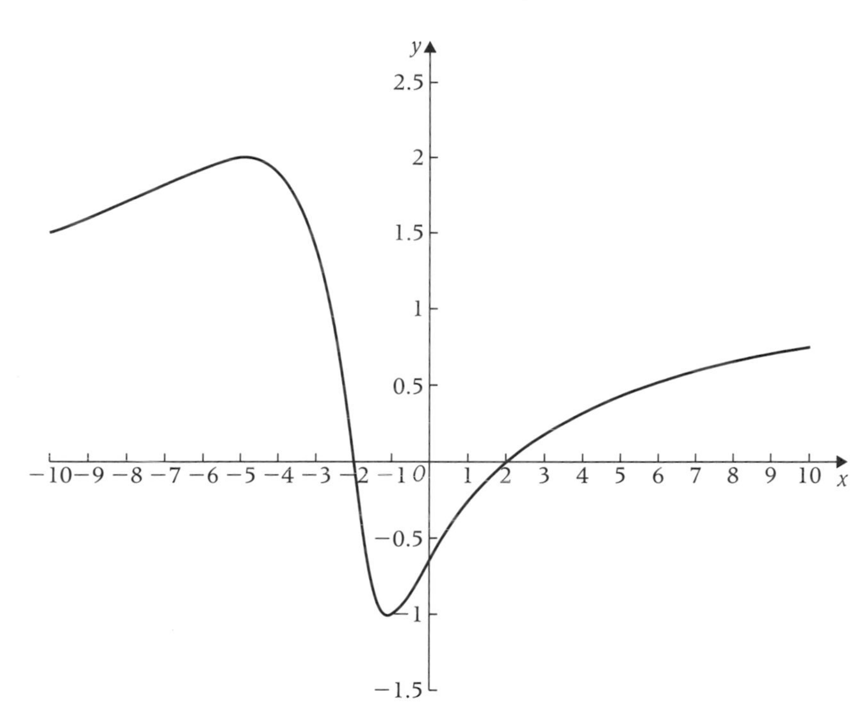

You can draw the curve on the screen and experiment by drawing lines such as $y = 3$, $y = 2$, $y = 1$, $y = 0$, $y = -1$, etc. to see which ones intersect the curve.

You should find that the lines $y = 2$ and $y = -1$ are tangential to the curve and you can verify that the curve only exists for $-1 \leqslant y \leqslant 2$.

12.4 Finding stationary points without calculus

In the previous section, you considered the curve $y = \dfrac{x + 5}{x^2 - x + 6}$ and showed that it only existed for values of y in the interval $-\frac{1}{23} \leqslant y \leqslant 1$. This means that if horizontal lines with equation $y = k$ are drawn to intersect the curve, they will only intersect for $-\frac{1}{23} \leqslant k \leqslant 1$.

The curve has no vertical asymptotes and so it is continuous and yet it is constrained to a narrow band of values. This means that the curve must reach a maximum point when $y = 1$.

When $\dfrac{x + 5}{x^2 - x + 6} = 1$, you can rearrange the equation to give

$$x + 5 = x^2 - x + 6 \quad \text{or} \quad x^2 - 2x + 1 = 0$$

$$\Rightarrow \quad (x - 1)^2 = 0,$$

which has a repeated root when $x = 1$.

The point $(1, 1)$ is a maximum point of the curve.

The situation for various lines of the form $y = k$ $(k > 0)$ is shown in the diagram.

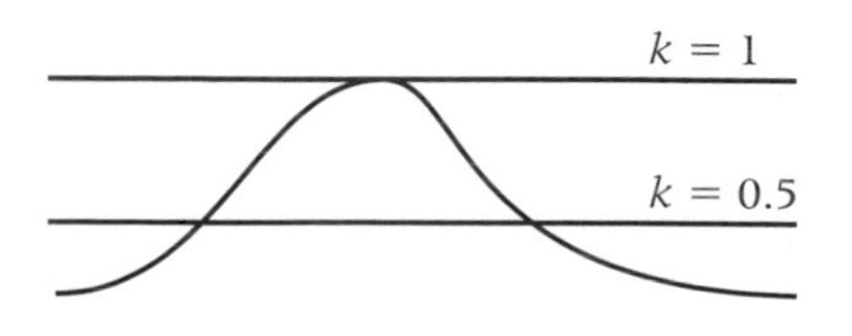

When $k = 1$, the line touches the curve and hence the quadratic has equal roots.

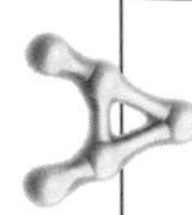

If a curve of the form $y = \dfrac{ax^2 + bx + c}{dx^2 + ex + f}$ is shown to exist only for $P \leqslant y \leqslant Q$, then it means that the curve must have a minimum point when $y = P$ and a maximum point when $y = Q$.

Substitute $y = P$ in order to find the x-coordinate of the minimum point. The resulting quadratic in x will always have a repeated root.

Repeat by substituting $y = Q$ to find the x-coordinate of the maximum point.

Worked example 12.4

A curve has equation $y = \dfrac{x^2 + 2}{x^2 - 4x}$.

(a) Find the equations of its asymptotes.

(b) Prove that no part of the curve exists for $-1 < y < \frac{1}{2}$.

(c) Hence find the coordinates of its stationary points.

(d) Sketch the curve.

Solution

(a) The denominator is zero when $x^2 - 4x = 0 \quad \Rightarrow \quad x(x-4) = 0$.
The vertical asymptotes have equations $x = 0$ and $x = 4$.
The horizontal asymptote can be found by dividing both the denominator and numerator by x^2.

$$y = \frac{x^2 + 2}{x^2 - 4x} = \frac{1 + \dfrac{2}{x^2}}{1 - \dfrac{4}{x}}$$

and since $\dfrac{1}{x} \to 0$ and $\dfrac{1}{x^2} \to 0$ when $|x| \to \infty$ it follows that $y \to 1$.

Hence the horizontal asymptote has equation $y = 1$.

(b) Consider $y = \dfrac{x^2 + 2}{x^2 - 4x} \quad \Rightarrow \quad y(x^2 - 4x) = x^2 + 2$.

You can rearrange to give a quadratic in x.

Rearranging gives $\Rightarrow \quad (y-1)x^2 - 4yx - 2 = 0$.

Since you are looking for places where no curve exists, you need the quadratic to have complex roots.

The discriminant must be negative, or $b^2 - 4ac < 0$.

$$(-4y)^2 - 4(y-1)(-2) < 0$$

or $\quad 16y^2 + 8y - 8 < 0$

You could have proceeded as before by using the line $y = k$ and this would have given you $16k^2 + 8k - 8 < 0$, etc.

Hence $2y^2 + y - 1 < 0 \Rightarrow (2y - 1)(y + 1) < 0$

A sign diagram for $(2y - 1)(y + 1)$ is shown opposite.

Hence $(2y - 1)(y + 1) < 0 \Rightarrow -1 < y < \frac{1}{2}$.

No part of the curve exists when y takes a value in this interval.

Note that you could have used the condition for real roots, namely:

$$(-4y)^2 - 4(y - 1)(-2) \geqslant 0$$

and concluded that the curve only existed for $y \geqslant \frac{1}{2}$ or $y \leqslant -1$.

Hence no part of the curve exists for $-1 < y < \frac{1}{2}$.

(c) The extreme values of y for which the curve exists must be when $y = -1$ and when $y = \frac{1}{2}$.

When $y = -1$, $\dfrac{x^2 + 2}{x^2 - 4x} = -1 \Rightarrow -(x^2 - 4x) = x^2 + 2$

$\Rightarrow 2x^2 - 4x + 2 = 0 \Rightarrow x^2 - 2x + 1 = 0$

$\Rightarrow (x - 1)^2 = 0$

If you have found the correct values of y, the quadratic in x will have equal roots.

Hence $x = 1$ (repeated).

One of the stationary points is therefore $(1, -1)$.

When $y = \dfrac{1}{2}$, $\dfrac{x^2 + 2}{x^2 - 4x} = \dfrac{1}{2} \Rightarrow \dfrac{1}{2}(x^2 - 4x) = x^2 + 2$

$\Rightarrow x^2 - 4x = 2x^2 + 4 \Rightarrow x^2 + 4x + 4 = 0$

$\Rightarrow (x + 2)^2 = 0$

Hence $x = -2$ (repeated).

The other stationary point is therefore $\left(-2, \dfrac{1}{2}\right)$.

(d) First draw in the asymptotes (remember $x = 0$ is an asymptote for this curve) and consider the behaviour near the vertical asymptotes.
Next, indicate the position of each of the stationary points.

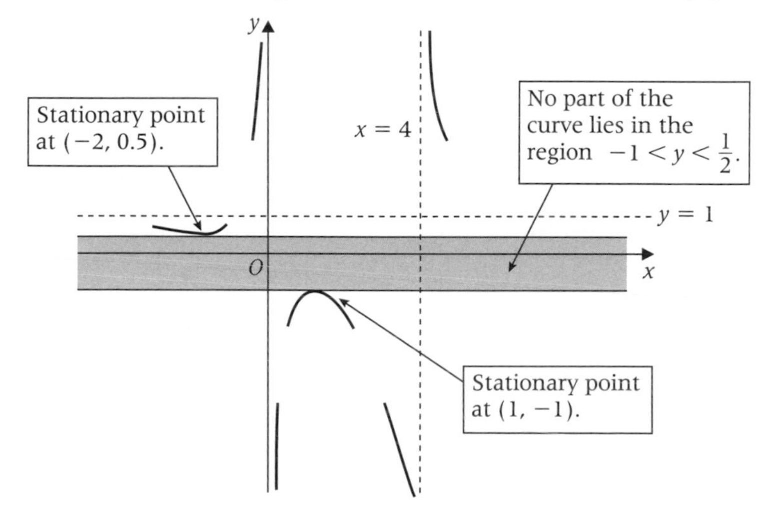

You can now draw the entire curve, realising that the curve must approach $y = 1$ for large positive and negative values of x.

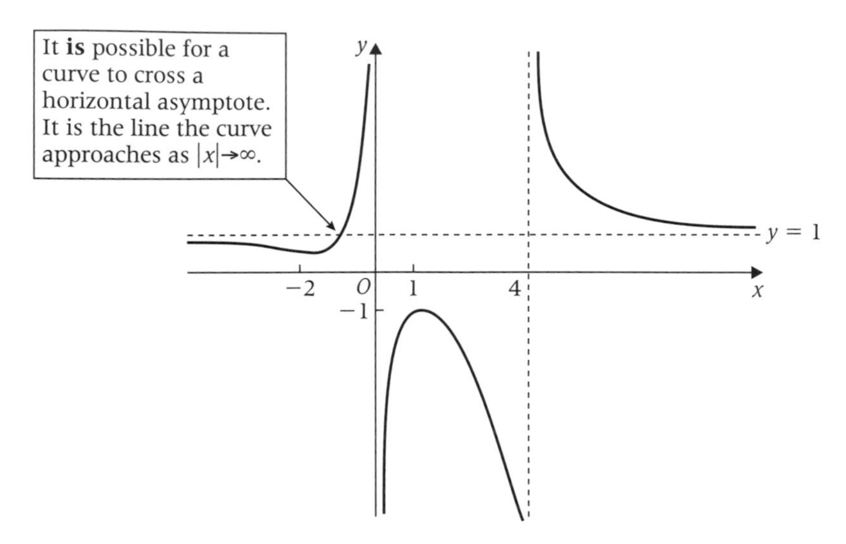

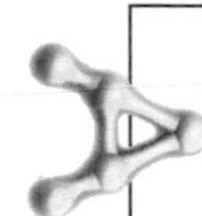

If a curve of the form $y = \dfrac{ax^2 + bx + c}{dx^2 + ex + f}$ is shown to exist only for $y \leqslant M$ or $y \geqslant N$, then it means that the curve must have a maximum point when $y = M$ and a minimum point when $y = N$.

Substitute $y = M$ in order to find the x-coordinate of the maximum point. The resulting quadratic in x will always have a repeated root.

Repeat by substituting $y = N$ to find the x-coordinate of the minimum point.

Worked example 12.5

The function f is defined by $\mathrm{f}(x) = \dfrac{12x^2 + 5x}{x^2 + 1}$.

(a) Write down the equation of the asymptote of the curve $y = \mathrm{f}(x)$.

(b) Show that the equation $\mathrm{f}(x) = k$ can be written in the form $(12 - k)x^2 + 5x - k = 0$.

(c) Hence find the values of k for which the equation $\mathrm{f}(x) = k$ has two equal roots.

(d) Deduce the coordinates of the stationary points of the curve with equation $y = \dfrac{12x^2 + 5x}{x^2 + 1}$.

Solution

(a) There are no vertical asymptotes since the equation $x^2 + 1 = 0$ has no real solutions.
As $x \to \infty$, $y \to 12$. Hence the curve has $y = 12$ as an asymptote.

(b) $\dfrac{12x^2 + 5x}{x^2 + 1} = k$ can be written in the form $12x^2 + 5x = k(x^2 + 1)$.
Hence $(12 - k)x^2 + 5x - k = 0$.

(c) For equal roots, the discriminant $(b^2 - 4ac)$ is zero.

$$5^2 - 4(12 - k)(-k) = 0$$

or $\quad 25 + 48k - 4k^2 = 0$

Hence $(25 - 2k)(1 + 2k) = 0$, so that $k = -\frac{1}{2}$ or $\frac{25}{2}$.

You could use the formula or complete the square if you prefer.

(d) The y-coordinates of the stationary points are given by the values of k found in **(c)**.
When $k = -\frac{1}{2}$, the quadratic in **(b)** becomes
$12.5x^2 + 5x + 0.5 = 0$

or $\quad 25x^2 + 10x + 1 = 0 \quad$ or $\quad (5x + 1)^2 = 0$.

Since the repeated root is $x = -0.2$, one stationary point has coordinates $(-0.2, -0.5)$.
When $k = \frac{25}{2}$, the quadratic in **(b)** becomes
$-0.5x^2 + 5x - 12.5 = 0$

or $\quad x^2 - 10x + 25 = 0 \quad$ or $\quad (x - 5)^2 = 0$.

Since the repeated root is $x = 5$, the other stationary point has coordinates $(5, 12.5)$.

EXERCISE 12B

1 A curve has equation $y = \dfrac{x^2 + 1}{x(4 + 3x)}$.

(a) Write down the equations of its asymptotes.

(b) Show that the curve intersects the line $y = k$ when $(3k - 1)x^2 + 4kx - 1 = 0$.

(c) Hence show that the line $y = k$ is a tangent to the curve when $4k^2 + 3k - 1 = 0$.

(d) Hence find the coordinates of the stationary points of the curve.

2 A curve has equation $y = \dfrac{x^2 - 6}{x^2 + 4x + 5}$.

(a) Prove that the curve only exists for $-3 \leqslant y \leqslant 2$.

(b) Hence determine the coordinates of the stationary points of the curve.

3 Show that the curve $y = \dfrac{x^2 + 2x + 2}{x^2 - x - 1}$ exists only when $y \leqslant -2$ and for $y \geqslant 0.4$. Hence find the coordinates of the stationary points of the curve.

4 A curve has equation $y = \dfrac{x + 1}{x(x - 3)}$.

(a) Show that there are no values of x for which $-1 < y < -\frac{1}{9}$.

(b) Hence determine the coordinates of any stationary points of the curve.

(c) State the equations of any asymptotes of the curve.

5 A curve has equation $y = \dfrac{x^2}{x^2 + 3x + 3}$.

(a) Prove that, for all real values of x, y satisfies the inequality $0 \leqslant y \leqslant 4$.

(b) Hence find the coordinates of the stationary points of the curve.

(c) Find the equations of any asymptotes of the curve.

6 A curve has equation $y = \dfrac{x + 2}{3 - x^2}$.

(a) Find the equations of any asymptotes and state the coordinate of any points where the curve crosses the coordinate axes.

(b) Determine the possible values y can take.

(c) Hence find the coordinates of the stationary points of the curve.

7 A curve has equation $y = \dfrac{2x - 3}{2x^2 - x - 1}$.

(a) Prove algebraically that no values of y exist in the interval $\frac{2}{9} < y < 2$.

(b) Hence find the coordinates of the stationary points of the curve.

(c) Find the equations of its three asymptotes and sketch the curve.

8 A curve has equation $y = \dfrac{x^2 - 4x - 5}{x^2 + x + 2}$.

(a) Prove algebraically that, for all real values of x, y satisfies the inequality $-\frac{18}{7} \leqslant y \leqslant 2$.

(b) Hence find the coordinates of the stationary points of the curve.

(c) Find the equations of any asymptotes and sketch the curve.

9 **(a)** Find the set of values of k for which the equation $k(x^2 + 4x + 9) = 10x$ has real roots.

(b) Hence determine the region for which the curve with equation $y = \dfrac{10x}{x^2 + 4x + 9}$ exists.

(c) Hence find the stationary points of $y = \dfrac{10x}{x^2 + 4x + 9}$ and sketch its graph.

10 A curve has equation $y = \dfrac{x^2 - x + 1}{x^2 + x + 1}$.

(a) Prove algebraically that $\frac{1}{3} \leqslant y \leqslant 3$.

(b) Use the result from **(a)** to find the coordinates of the maximum and minimum points of the curve.

(c) Show that there are no vertical asymptotes and state the equation of the horizontal asymptote of the curve.

(d) Sketch the curve.

Key point summary

1 It is always useful to draw any asymptotes as the first stage in sketching the graph of a rational function. *p159*

2 When a curve has a vertical asymptote at $x = a$, it is useful to check the values of y when x is a little smaller than a and when x is a little larger than a. *p159*

By considering the behaviour very close to the asymptotes, it is often possible to deduce the main shape of the graph.

3 In order to find the set of values of y for which a curve of the form $y = \dfrac{ax^2 + bx + c}{dx^2 + ex + f}$ exists: *p162*

1 consider where $y = k$ cuts the curve by writing $k = \dfrac{ax^2 + bx + c}{dx^2 + ex + f}$;

2 multiply out to obtain a quadratic of the form $Ax^2 + Bx + C = 0$, where A, B and C will involve k;

3 use the condition for real roots $B^2 - 4AC \geqslant 0$ to obtain a quadratic inequality involving k,

4 convert the solution involving k to a condition involving y.

4 If a curve of the form $y = \dfrac{ax^2 + bx + c}{dx^2 + ex + f}$ is shown to exist only for $P \leqslant y \leqslant Q$, then it means that the curve must have a minimum point when $y = P$ and a maximum point when $y = Q$. p164

Substitute $y = P$ in order to find the x-coordinate of the minimum point. The resulting quadratic in x will always have a repeated root.

Repeat by substituting $y = Q$ to find the x-coordinate of the maximum point.

5 If a curve of the form $y = \dfrac{ax^2 + bx + c}{dx^2 + ex + f}$ is shown to exist only for $y \leqslant M$ or $y \geqslant N$, then it means that the curve must have a maximum point when $y = M$ and a minimum point when $y = N$. p166

Substitute $y = M$ in order to find the x-coordinate of the maximum point. The resulting quadratic in x will always have a repeated root.

Repeat by substituting $y = N$ to find the x-coordinate of the minimum point.

Test yourself | What to review

1 A curve has equation $y = \dfrac{3}{(x-1)(x+2)}$. Section 12.2

(a) State the equations of its asymptotes.

(b) Sketch the curve.

2 A curve has equation $y = \dfrac{4x}{x^2 + 4}$. Section 12.4

(a) Write down the equations of any asymptotes

(b) (i) Show that the curve intersects the line $y = k$ when $kx^2 - 4x + 4k = 0$.

(ii) Find the values of k for which the quadratic in **(i)** has equal roots.

(c) Hence find the coordinates of the stationary points of the curve.

3 A curve has equation $y = \dfrac{x^2 - 2x + 1}{x^2 + x + 1}$. Sections 12.3 and 12.4

(a) Prove that, for all real values of x, the value of y satisfies the inequality $0 \leqslant y \leqslant 4$.

(b) Hence find the coordinates of the stationary points of the curve.

Test yourself ANSWERS

1 (a) $x = 1, \quad x = -2, \quad y = 0;$

(b)

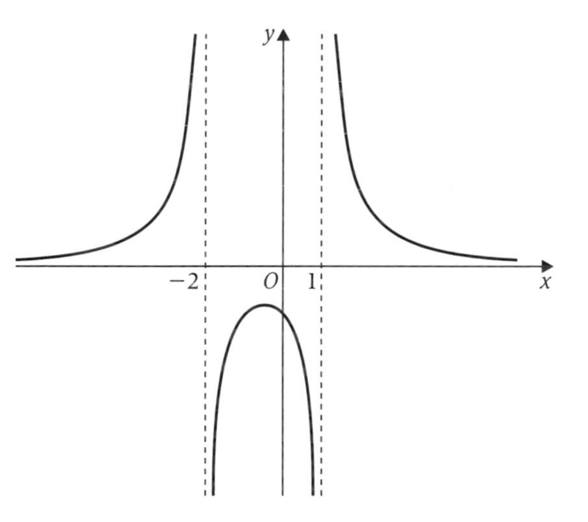

2 (a) $y = 0;$ **(b) (ii)** $1, -1;$ **(c)** $(2, 1), (-2, -1).$

3 (b) $(-1, 4), \quad (1, 0).$

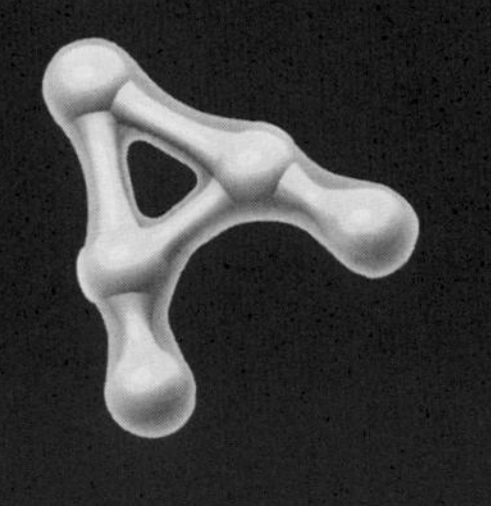

CHAPTER 13

Parabolas, ellipses and hyperbolas

Learning objectives

After studying this chapter, you should be able to:

- recognise standard equations for parabolas, ellipses and hyperbolas and sketch their graphs
- transform these curves by reflecting in the line $y = x$
- use translations to transform curves
- transform curves by stretches parallel to the coordinate axes
- find points of intersection of these graphs with coordinate axes and other straight lines
- interpret the geometrical implication of equal roots, distinct real roots or no real roots.

13.1 Parabolas with their vertices at the origin

You have already seen (C1, chapter 4) how to draw a parabola with equation $y = x^2$. It has the familiar U shape with its vertex at the origin.

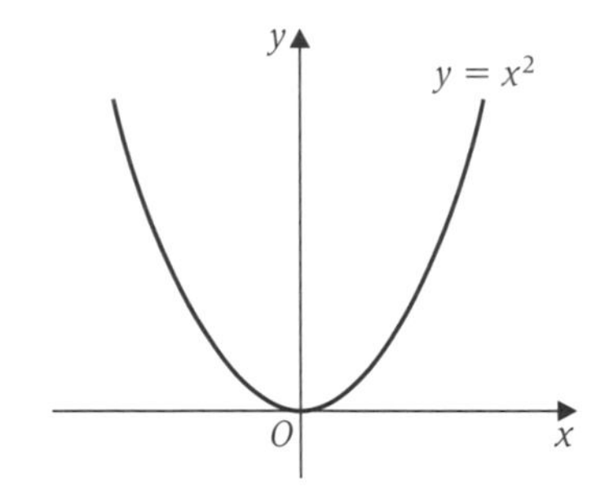

Suppose you reflect the curve in the line $y = x$. This will interchange the variables x and y. Hence the graph of $y = x^2$ is transformed into $x = y^2$ or, as it is more commonly written, $y^2 = x$.

Its graph is drawn below.

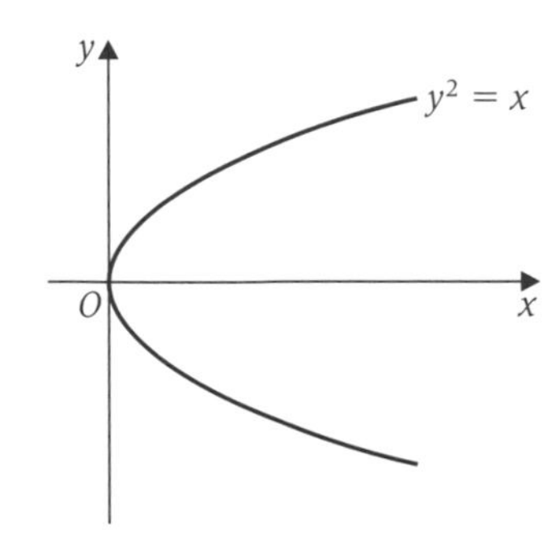

The vertex is still at the origin after the reflection.

A reflection in the line $y = x$ maps (x, y) onto (x', y'), where $x' = y$. The equation of the new curve after a reflection in $y' = x$
the line $y = x$ is obtained by interchanging x and y in the original equation.

Worked example 13.1

Find the equation of the new curve obtained by reflecting the curve with equation $3y = (x - 5)^2$ in the line $y = x$.

Sketch the curve with equation $3y = (x - 5)^2$ and also sketch its image after being reflected in the line $y = x$.

Solution

Interchange the variables x and y in the equation of the original curve.

You will now get $3x = (y - 5)^2$.

You may recognise the original curve as a parabola with vertex at $(5, 0)$.

The new curve will also be a parabola. Its vertex is at $(0, 5)$.

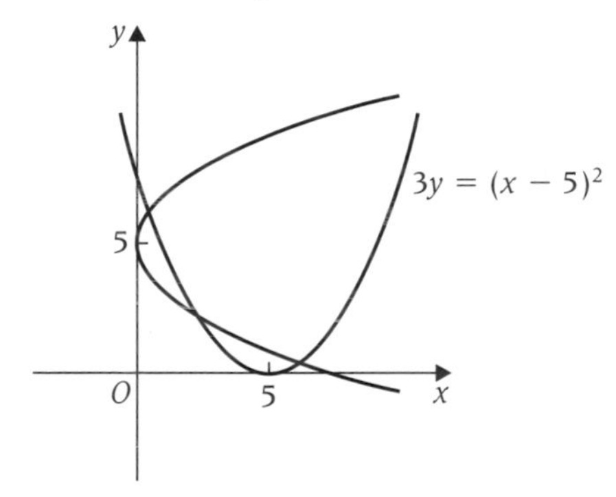

A parabola with its vertex at the origin and its axis along the x-axis will have an equation of the form $y^2 = kx$, where k is a constant.

13.2 Ellipses with their centres at the origin

You are already familiar with the equation of a circle of the form $x^2 + y^2 = 16$. It has radius 4 and its centre is at the origin. It cuts the x-axis when $y = 0$. This occurs when $x^2 = 16 \quad \Rightarrow \quad x = \pm 4$. Similarly, it cuts the y-axis when $y^2 = 16 \quad \Rightarrow \quad y = \pm 4$.

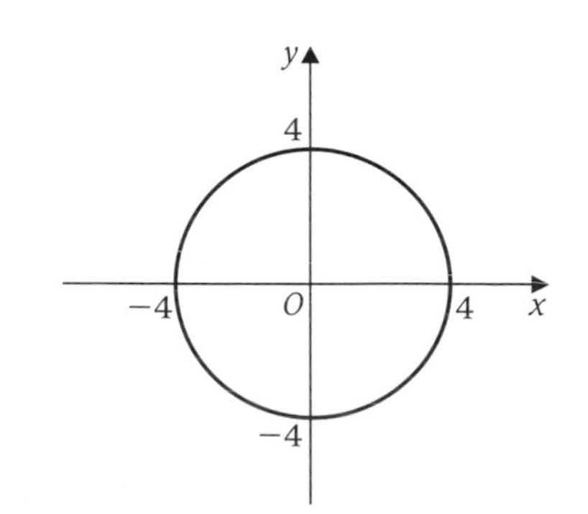

Suppose you were to stretch the circle in the x-direction with scale factor 1.5. The new curve would still cut the y-axis when $y = \pm 4$, but it would cut the x-axis when $x = \pm 6$.
The new shape is called an ellipse.

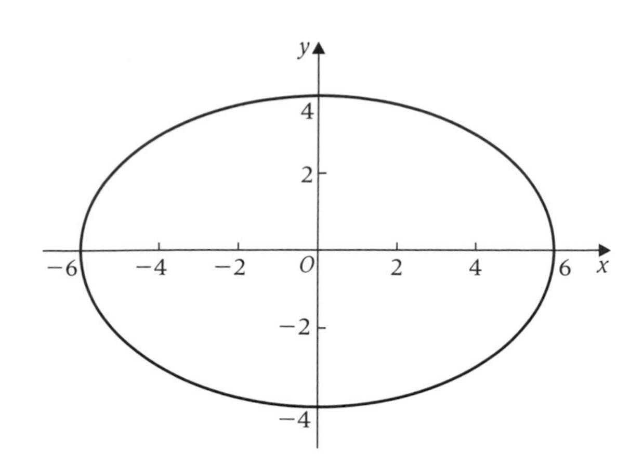

The ellipse will have equation $\left(\frac{2x}{3}\right)^2 + y^2 = 16$ or $4x^2 + 9y^2 = 144$.

You can see from this equation that when $y = 0$,
$4x^2 = 144 \quad \Rightarrow \quad x^2 = 36 \quad \Rightarrow \quad x = \pm 6$.

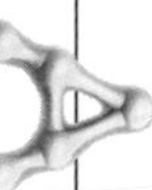

A stretch of scale factor c in the x-direction and scale factor d in the y-direction, maps (x, y) onto (x', y'), where $\begin{matrix} x' = cx \\ y' = dy \end{matrix}$.

The equation of the new curve is obtained by replacing x by $\left(\frac{x}{c}\right)$ and y by $\left(\frac{y}{d}\right)$ in the original equation.

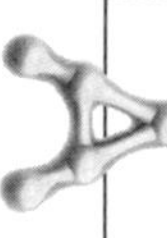

The general equation of an ellipse with its centre at the origin is $\frac{x^2}{a^2} + \frac{y^2}{b^2} = 1$.

It cuts the x-axis when $x = \pm a$ and cuts the y-axis when $y = \pm b$.

The circle with centre (0, 0) and radius 1 has equation $x^2 + y^2 = 1$. By applying a stretch of scale factor a in the x-direction and scale factor b in the y-direction you obtain the ellipse below.

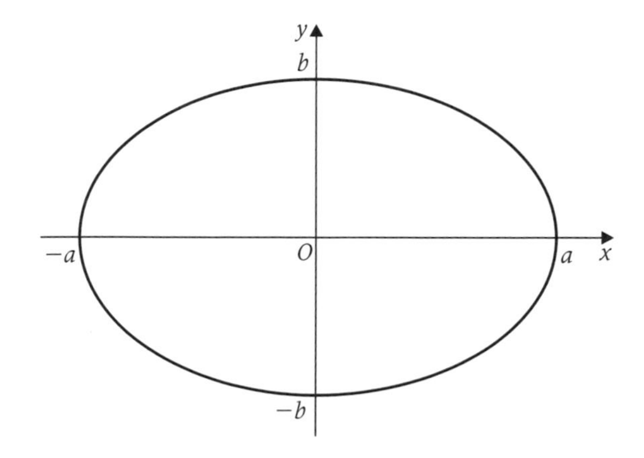

The diagram illustrates the case when $a > b$ and in this case the x-axis is said to be the major axis and the y-axis is said to be the minor axis.

The ellipse is said to have semi-major axis of length a and semi-minor axis of length b.

Worked example 13.2

The ellipse with equation $4x^2 + 9y^2 = 144$ is reflected in the line $y = x$. Find the equation of the new curve and sketch its graph.

Solution

The equation of the new curve is found by interchanging the variables x and y to give $4y^2 + 9x^2 = 144$.

$$\Rightarrow \quad 9x^2 + 4y^2 = 144 \quad \text{or} \quad \frac{x^2}{16} + \frac{y^2}{36} = 1$$

This is an ellipse with semi-axes 4 and 6. The major axis this time is in the y-direction.
Its graph is sketched below:

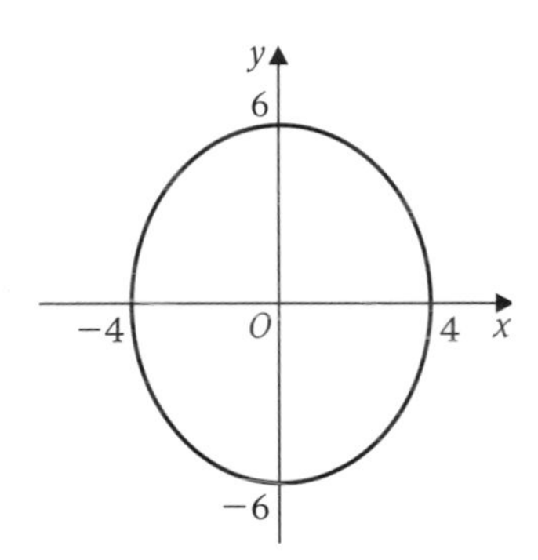

EXERCISE 13A

1 Find the equation of the curve obtained by reflecting the curve with equation $y = (x + 3)^2$ in the line $y = x$.
Sketch the curve with equation $y = (x + 3)^2$ and also sketch its image after being reflected in the line $y = x$.

2 The curve C_1 has equation $y = 4x^2 - 1$.

(a) Find the equation of the curve C_2 produced by reflecting C_1 in the line $y = x$. Sketch the curves C_1 and C_2 on the same axes.

(b) Find the equation of the curve C_3 produced by stretching C_1 by scale factor 2 in the x-direction.

(c) Find the equation of the curve C_4 produced by stretching C_1 by scale factor $\frac{1}{2}$ in the y-direction.

3 The ellipse with equation $x^2 + 4y^2 = 16$ is reflected in the line $y = x$.

(a) Find the equation of the new curve and sketch its graph.

(b) State the coordinates of the points where the new curve cuts the coordinate axes.

4 Find the equation of the curve produced when the circle with centre the origin and radius 3 is stretched by a factor of 2 in the x-direction.
Sketch the curve, marking clearly any values where the curve crosses the coordinate axes.

5 State clearly the geometrical transformations involved in transforming:

(a) the curve $y = 2x^2$ into the curve $y^2 = 2x$,

(b) the curve $y = 2(x^2 - 3)$ into the curve $y = 8(x^2 - 3)$,

(c) the curve $y = 4(x^2 + 1)$ into the curve $y = (x^2 + 4)$.

6 State clearly the geometrical transformations involved in transforming:

(a) the curve $9x^2 + y^2 = 1$ into the curve $x^2 + y^2 = 1$,

(b) the curve $3x^2 + 4y^2 = 12$ into the curve $4x^2 + 3y^2 = 12$,

(c) the curve $3x^2 + 4y^2 = 12$ into the curve $3x^2 + y^2 = 12$.

13.3 Hyperbolas

The curve with equation $xy = 2$ may seem a little unfamiliar in this form. However, once it is rearranged in the form $y = \frac{2}{x}$, you should identify it immediately as the hyperbola sketched below.

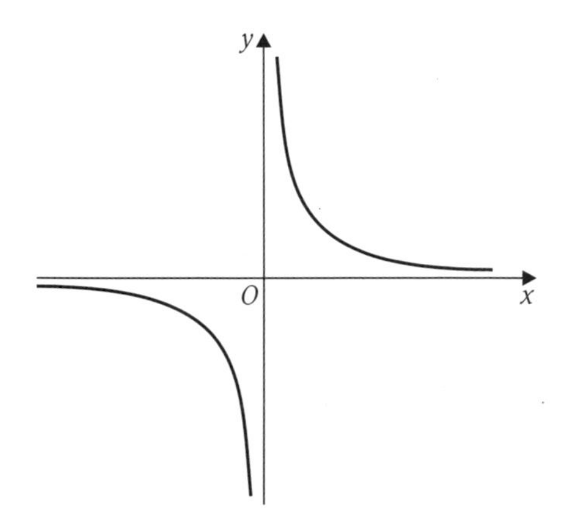

Clearly the graph is identical if you reflect it in the line $y = x$. This is confirmed by interchanging x and y to give

$$x = \frac{2}{y} \quad \Rightarrow \quad y = \frac{2}{x}.$$

Worked example 13.3

Find the equation of the new curve produced when the curve with equation $xy = 9$ is transformed by a stretch of scale factor 4 in the x-direction.

Solution

To find the new equation, replace x by $\left(\frac{x}{4}\right)$.

The new curve has the equation $\left(\frac{x}{4}\right)y = 9 \quad \Rightarrow \quad xy = 36.$

Rotation through 45°

In chapter 6, you learned to use matrices for transformations.

Suppose you wish to rotate the point (a, b) through 45° to reach a new point (p, q).

$$\begin{bmatrix} p \\ q \end{bmatrix} = \begin{bmatrix} \cos 45° & -\sin 45° \\ \sin 45° & \cos 45° \end{bmatrix} \begin{bmatrix} a \\ b \end{bmatrix}$$

$$\Rightarrow \quad \begin{bmatrix} p \\ q \end{bmatrix} = \begin{bmatrix} \frac{1}{\sqrt{2}} & -\frac{1}{\sqrt{2}} \\ \frac{1}{\sqrt{2}} & \frac{1}{\sqrt{2}} \end{bmatrix} \begin{bmatrix} a \\ b \end{bmatrix}$$

Hence $p = \frac{1}{\sqrt{2}}(a - b)$ and $q = \frac{1}{\sqrt{2}}(a + b)$.

Multiplying together gives $pq = \frac{1}{2}(a^2 - b^2)$.

This suggests that if the point (a, b) lies on the curve with equation $x^2 - y^2 = 2$, then the point (p, q) lies on the hyperbola $xy = 1$.

Hence the curve with equation $x^2 - y^2 = 2$ must also represent a hyperbola since only a rotation through 45° has taken place in moving from one graph to the other.

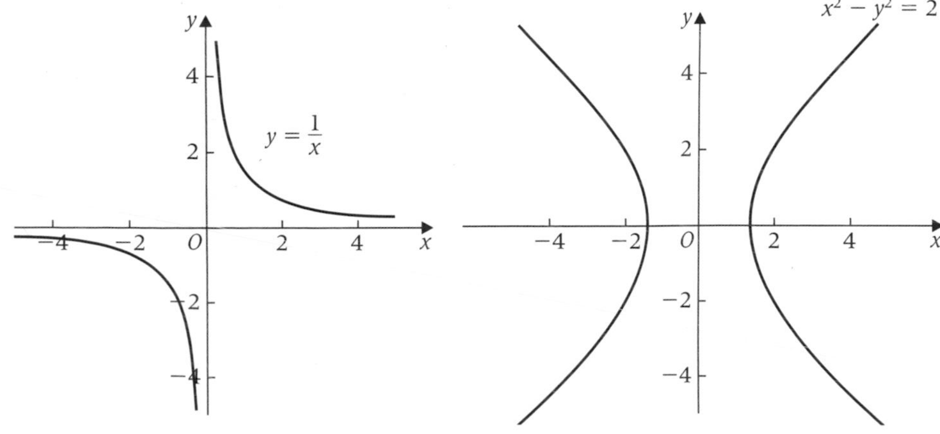

It should be clear that by appropriate stretches in the x- and y-directions the curves with equations $x^2 - 3y^2 = 4$ and $\frac{x^2}{25} - \frac{y^2}{9} = 1$, etc. are also hyperbolas.

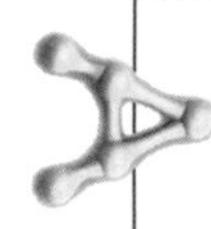

An equation of the form $xy = c^2$ represents a hyperbola with the coordinate axes as asymptotes. The asymptotes are perpendicular and it is often referred to as a rectangular hyperbola.

It can be rotated through 45° to give an equation of the form $x^2 - y^2 = k$.

Worked example 13.4

A hyperbola, H_1, has equation $\frac{x^2}{9} - y^2 = 1$.

(a) Find the points where H_1 intersects the coordinate axes.

(b) Find the equation of the curve H_2 formed by reflecting H_1 in the line $y = x$. Sketch H_2.

Solution

(a) H_1 cuts the x-axis when $y = 0$.

Hence $\frac{x^2}{9} = 1 \quad \Rightarrow \quad x = \pm 3$.

The points on the x-axis are $(3, 0)$ and $(-3, 0)$.

H_1 cuts the y-axis when $x = 0$.

This would mean $-y^2 = 1$, which is impossible.
There are no points of intersection with the y-axis.

(b) The curve H_2 has equation $\frac{y^2}{9} - x^2 = 1$, which is obtained by interchanging the variables x and y. It is another hyperbola.

H_2 will intersect the y-axis at $(0, 3)$ and $(0, -3)$ but will not cut the x-axis at all.

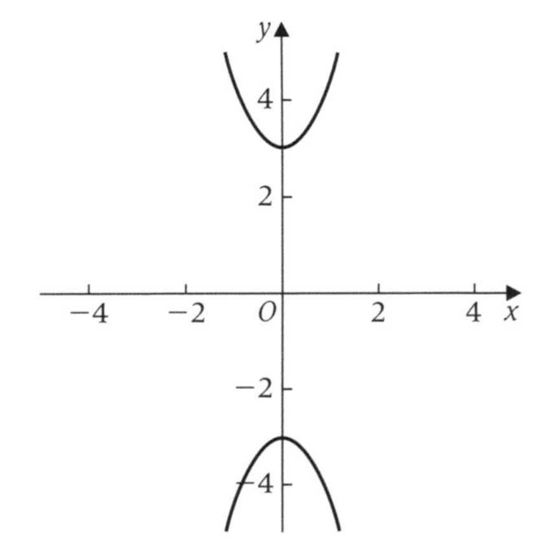

When x and y are numerically large $\frac{y^2}{9} \approx x^2 \quad \Rightarrow \quad y = \pm 3x$ are the equations of the asymptotes.

The equation $\frac{x^2}{a^2} - \frac{y^2}{b^2} = 1$ represents a hyperbola with centre at the origin cutting the x-axis when $x = \pm a$.
It does not intersect the y-axis and its asymptotes have equations $y = \pm\frac{b}{a}x$. When $b = a$ the hyperbola is said to be a rectangular hyperbola.

EXERCISE 13B

1 Find the equation of the hyperbola obtained by reflecting the hyperbola with equation $x^2 - y^2 = 25$ in the line $y = x$.
Sketch the two hyperbolas on the same axes.

2 The hyperbola H_1 has equation $\frac{x^2}{4} - \frac{y^2}{9} = 1$.

(a) Find the equation of the curve H_2 produced by reflecting H_1 in the line $y = x$. Sketch the curves H_1 and H_2 on the same axes.

(b) Find the equation of the curve H_3 produced by stretching H_1 by scale factor 2 in the y-direction.

(c) Find the equation of the curve H_4 produced by stretching H_1 by scale factor $\frac{1}{2}$ in the x-direction.

3 The hyperbola with equation $x^2 - 4y^2 = 16$ is reflected in the line $y = x$.

(a) Find the equation of the new curve and sketch its graph.

(b) State the coordinates of the points where the new curve cuts the coordinate axes.

4 State clearly the geometrical transformations involved in transforming:

(a) the curve $9x^2 - y^2 = 1$ into the curve $x^2 - y^2 = 1$,

(b) the curve $3x^2 - 4y^2 = 12$ into the curve $4x^2 - 3y^2 = -12$,

(c) the curve $xy = 7$ into the curve $xy = 21$,

(d) the curve $3x^2 - 4y^2 = 12$ into the curve $x^2 - y^2 = 4$.

5 **(a)** The hyperbola $xy = 5$ is transformed by a stretch with scale factor 2 in the x-direction.

(i) Find the equation of the resulting curve.

(ii) Show that the transformation could have been achieved by a stretch with the same scale factor in the y-direction.

(b) The hyperbola $x^2 - y^2 = 10$ is transformed by a stretch with scale factor 2 in the x-direction.

(i) Find the equation of the resulting curve.

(ii) Can the transformation be achieved by a stretch with an appropriate scale factor in the y-direction?

13.4 Translations of curves

In C2 chapter 5 you learnt how to translate curves with equations of the form $y = f(x)$ through the vector $\begin{bmatrix} c \\ d \end{bmatrix}$ and obtained a curve with equation $y - d = f(x - c)$.

In effect, you replaced x by $x - c$ and replaced y by $y - d$.

Although, the parabolas, ellipses and hyperbolas considered above do not have equations precisely of this form, you can still use a similar procedure to find the new equations.

> A translation with vector $\begin{bmatrix} c \\ d \end{bmatrix}$ maps (x, y) onto (x', y'),
>
> where $\begin{aligned} x' &= x + c \\ y' &= y + d \end{aligned}$.
>
> The equation of the new curve is obtained by replacing x by $(x - c)$ and y by $(y - d)$ in the original equation.

Worked example 13.5

(a) The hyperbola with equation $\frac{x^2}{9} - \frac{y^2}{4} = 1$ is translated by 3 units in the x-direction and 5 units in the y-direction. Find the equation of the resulting curve.

(b) Find the equation of the curve formed by translating the parabola with equation $y^2 = 4x$ through the vector $\begin{bmatrix} -3 \\ 1 \end{bmatrix}$.

Solution

(a) You need to replace x by $(x - 3)$ and y by $(y - 5)$.

The new hyperbola has equation $\frac{(x-3)^2}{9} - \frac{(y-5)^2}{4} = 1$.

You are not expected to multiply out the brackets.

The new centre will have moved to the point $(3, 5)$.

(b) This time, you need to replace x by $(x + 3)$ and y by $(y - 1)$. The new equation is $(y - 1)^2 = 4(x + 3)$.

Worked example 13.6

Find the transformation that has taken place when the curve with equation $y^2 = 2x$ is mapped onto the curve with equation $y(y - 4) = 2(x - 3)$.

Hence, sketch the curve with equation $y(y - 4) = 2(x - 3)$, stating the coordinates of its vertex and the equation of its axis of symmetry.

Solution

You need to multiply out the final equation to give $y^2 - 4y = 2x - 6$.

Now, completing the square gives

$$y^2 - 4y + 4 = 2x - 6 + 4 = 2x - 2$$
$$(y - 2)^2 = 2\ (x - 1)$$

You can now see that this equation is the result of changing the variables in the equation of the original curve $y^2 = 2x$.

The variable x has been replaced by $(x - 1)$ and the variable y by $(y - 2)$.

The transformation is a translation with vector $\begin{bmatrix} 1 \\ 2 \end{bmatrix}$.

In order to sketch the curve, you need only consider the graph of $y^2 = 2x$ being translated through $\begin{bmatrix} 1 \\ 2 \end{bmatrix}$.

Its graph is sketched below. The parabola has vertex $(1, 2)$ and its axis of symmetry has equation $y = 2$.

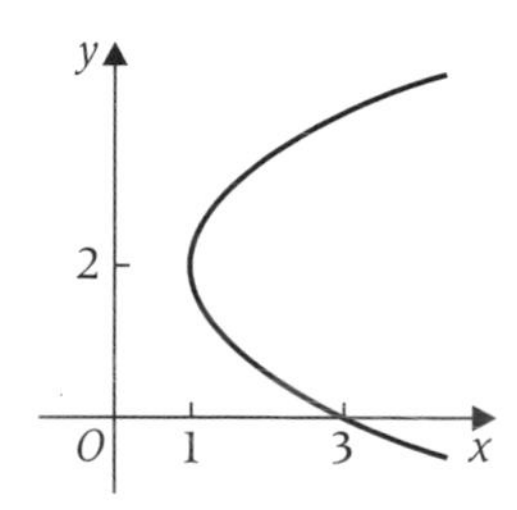

EXERCISE 13C

1 Find the equation of the curve resulting from translating the curve with equation $\frac{x^2}{2} + \frac{y^2}{3} = 1$ through $\begin{bmatrix} -3 \\ 2 \end{bmatrix}$.

2 The following curves are translated through the vectors indicated. Find the equations of the new curves.

(a) $\frac{x^2}{4} - \frac{y^2}{5} = 1$, $\begin{bmatrix} 3 \\ 0 \end{bmatrix}$,

(b) $5x^2 + 7y^2 = 12$, $\begin{bmatrix} -4 \\ -2 \end{bmatrix}$,

(c) $x^2 - y^2 = 3$, $\begin{bmatrix} 2 \\ -5 \end{bmatrix}$,

(d) $(x - 1)(y + 2) = 3$, $\begin{bmatrix} -1 \\ 2 \end{bmatrix}$,

(e) $y^2 = x - 3$, $\begin{bmatrix} -3 \\ -1 \end{bmatrix}$,

(f) $(x - 1)^2 - (y + 3)^2 = 7$, $\begin{bmatrix} 2 \\ 3 \end{bmatrix}$.

3 The curve with equation $xy = 4$ is transformed into the curve with equation $(x - 5)(y + 4) = 4$. Describe geometrically the transformation that has taken place.

4 The curve with equation $x^2 - y^2 = 1$ is translated onto the curve with equation $x^2 - 6x - y^2 + 2y = k$. Find the value of k and find the vector of the translation.

5 The curve with equation $y^2 = 6x$ is transformed onto the curve with equation $y^2 = 4y + 6(x + 1)$. Describe geometrically the transformation that has taken place.

6 The curve with equation $x^2 - 2x - y^2 + 4y = 20$ is translated onto the curve with equation $x^2 - 8x - y^2 - 2y = k$. Find the value of k and find the vector of the translation.

13.5 Intersections with straight lines

When you considered parabolas and circles in C1, you realised that a line may intersect in two distinct points or not intersect at all. Alternatively, when the quadratic being solved has a single repeated root, the line was shown to be a tangent to the parabola or circle being considered.

Worked example 13.7

Find any points of intersection of the line $y = 3x - 5$ and the parabola $y^2 = kx - 2$ in the cases where **(a)** $k = 6$, and **(b)** $k = -3$.

Solution

You need to substitute $y = 3x - 5$ into the equation $y^2 = kx - 2$.

$$(3x - 5)^2 = kx - 2 \quad \Rightarrow \quad 9x^2 - 30x + 25 = kx - 2$$
$$\Rightarrow \quad 9x^2 - (30 + k)x + 27 = 0$$

(a) When $k = 6$, the quadratic becomes $9x^2 - 36x + 27 = 0$ or $x^2 - 4x + 3 = 0 \quad \Rightarrow \quad (x - 1)(x - 3) = 0$.

Hence $x = 1$ or $x = 3$.

You can find the y-values by substituting into $y = 3x - 5$. The line intersects the parabola at $(1, -2)$ and $(3, 4)$.

(b) When $k = -3$, the quadratic becomes $9x^2 - 27x + 27 = 0$ or $x^2 - 3x + 3 = 0$.

The discriminant is $(-3)^2 - 12 = -3$ which is negative.

Hence there are no points of intersection when $k = -3$.

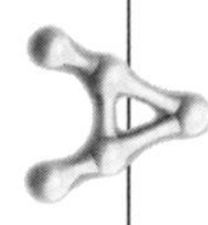

To consider how a straight line intersects a parabola, ellipse or hyperbola:

1 form a quadratic equation (usually in terms of x) of the form $ax^2 + bx + c = 0$;

2 when $b^2 - 4ac < 0$ there are no points of intersection;

3 when $b^2 - 4ac > 0$ there are two distinct points of intersection;

4 when $b^2 - 4ac = 0$ there is a single point of intersection. The line is a tangent to the curve at that point.

Worked example 13.8

Find the possible values of the constant m so that the line $y = mx + 3$ is a tangent to the ellipse $2x^2 + y^2 = 6$.

Solution

You need to substitute $y = mx + 3$ into the equation of the ellipse, namely $2x^2 + y^2 = 6$.

This gives $2x^2 + (mx + 3)^2 = 6$.

Hence $2x^2 + m^2x^2 + 6mx + 9 = 6$

or $(2 + m^2)x^2 + 6mx + 3 = 0$.

For the line to be a tangent, the quadratic must have equal roots. This means that the discriminant, $b^2 - 4ac = 0$.

$$36m^2 - 12(2 + m^2) = 0 \Rightarrow 3m^2 = 2 + m^2$$
$$\Rightarrow m^2 = 1 \Rightarrow m = \pm 1$$

EXERCISE 13D

1 Find any points of intersection of the line $y = kx - 5$ and the hyperbola $y^2 - x^2 = 8$ in the cases where: **(a)** $k = 2$, and **(b)** $k = 0$.

2 Prove that the line $y = 2x + 1$ is a tangent to the parabola $y^2 = 8x$.

3 **(a)** Determine whether the line with equation $x + 3y = 1$ intersects the ellipse with equation $x^2 + 3y^2 = 1$.

(b) Find the possible values of k for which the line with equation $x + 3y = k$ intersects the ellipse $x^2 + 3y^2 = 1$.

4 Find the possible values of m such that the line $y = mx - 7$ is a tangent to the hyperbola with equation $x^2 - 2y^2 = 14$.

5 Find any points of intersection of the line with equation $y = 3x - 2$ and the following curves:

(a) the parabola $y^2 = x$,

(b) the ellipse $\frac{x^2}{8} + \frac{y^2}{32} = 1$,

(c) the hyperbola $y^2 - 10x^2 = 15$.

6 Find the possible values of the constant p for which the line with equation $2x + y = p$ does not intersect the ellipse with equation $2(x - 1)^2 + y^2 = 3$.

7 **(a)** Determine whether the line with equation $2y = 3x - 7$ intersects the hyperbola with equation $x^2 - 2y^2 = 8$.

(b) Find the possible values of the constant k for which the line $2y = 3x - 7$ intersects the hyperbola $x^2 - 2y^2 = k$.

Key point summary

1 A reflection in the line $y = x$ maps (x, y) onto (x', y'), where $\begin{matrix} x' = y \\ y' = x \end{matrix}$. p173

The equation of the new curve after a reflection in the line $y = x$ is obtained by interchanging x and y in the original equation.

2 A parabola with its vertex at the origin and its axis along the x-axis will have an equation of the form $y^2 = kx$, where k is a constant. p173

3 A stretch of scale factor c in the x-direction and scale factor d in the y-direction, maps (x, y) onto (x', y'), where $\begin{matrix} x' = cx \\ y' = dy \end{matrix}$. p174

The equation of the new curve is obtained by replacing x by $\left(\frac{x}{c}\right)$ and y by $\left(\frac{y}{d}\right)$ in the original equation.

4 The general equation of an ellipse with its centre at the origin is $\frac{x^2}{a^2} + \frac{y^2}{b^2} = 1$. p174

It cuts the x-axis when $x = \pm a$ and cuts the y-axis when $y = \pm b$.

5 An equation of the form $xy = c^2$ represents a hyperbola with the coordinate axes as asymptotes. It is often referred to as a rectangular hyperbola. p178

It can be rotated through 45° to give an equation of the form $x^2 - y^2 = k$.

6 The equation $\frac{x^2}{a^2} - \frac{y^2}{b^2} = 1$ represents a hyperbola with centre at the origin cutting the x-axis when $x = \pm a$. p178

It does not intersect the y-axis and its asymptotes have equations $y = \pm\frac{b}{a}x$. When $b = a$ the hyperbola is said to be a rectangular hyperbola.

7 A translation with vector $\begin{bmatrix} c \\ d \end{bmatrix}$ maps (x, y) onto (x', y'), where $\begin{matrix} x' = x + c \\ y' = y + d \end{matrix}$. p180

The equation of the new curve is obtained by replacing x by $(x - c)$ and y by $(y - d)$ in the original equation.

8 To consider how a straight line intersects a parabola, ellipse or hyperbola: *p182*

1 form a quadratic equation (usually in terms of x) of the form $ax^2 + bx + c = 0$;

2 when $b^2 - 4ac < 0$ there are no points of intersection;

3 when $b^2 - 4ac > 0$ there are two distinct points of intersection;

4 when $b^2 - 4ac = 0$ there is a single point of intersection. The line is a tangent to the curve at that point.

Test yourself	What to review
1 Determine whether the following equations represent a circle, an ellipse, a parabola or a hyperbola: **(a)** $y^2 = 6x$, **(b)** $x^2 - 2y^2 = 1$, **(c)** $3x^2 + 2y^2 = 12$.	*Sections 13.1, 13.2 and 13.3*
2 The curve C has equation $4x^2 + 9y^2 = 36$. **(a)** Find the points where C intersects the coordinate axes. **(b)** Find the equation of the curve when C is reflected in the line $y = x$.	*Sections 13.1*
3 Find the new curve formed when the curve with equation $x^2 - 9y^2 = 1$ is stretched by a scale factor 3 in the y-direction.	*Sections 13.2*
4 The parabola with equation $(y + 3)^2 = 6x - 4$ is translated by the vector $\begin{bmatrix} -1 \\ 3 \end{bmatrix}$. Find the equation of the new curve.	*Sections 13.4*
5 Find the possible values of the constant m for which the line $y = mx + 10$ is a tangent to the ellipse with equation $4x^2 + y^2 = 20$.	*Sections 13.5*

Test yourself ANSWERS

1 **(a)** parabola; **(b)** hyperbola; **(c)** ellipse.

2 **(a)** $(3, 0)$, $(-3, 0)$, $(0, 2)$, $(0, -2)$; **(b)** $9x^2 + 4y^2 = 36$.

3 $x^2 - y^2 = 1$.

4 $y^2 = 6x + 2$.

5 $m = \pm 4$.

Exam style practice paper

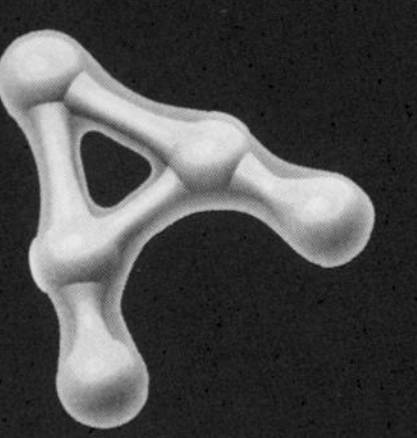

Time allowed 1 hour 30 minutes

Answer **all** questions

---

1 The roots of the quadratic equation $x^2 - 3x + 4 = 0$ are α and β.

(a) Without solving the equation,

(i) write down the value of $\alpha + \beta$ and the value of $\alpha\beta$, (2 marks)

(ii) show that $\alpha^3 + \beta^3 = -9$, (3 marks)

(iii) find the value of $\dfrac{2\alpha}{\beta^2} + \dfrac{2\beta}{\alpha^2}$. (2 marks)

(b) Determine a quadratic equation with integer coefficients which has roots $\dfrac{2\alpha}{\beta^2}$ and $\dfrac{2\beta}{\alpha^2}$. (3 marks)

2 (a) Find the complex roots of the equation $x^2 + 4x + 13 = 0$. (4 marks)

(b) Find the complex number z that satisfies the equation $z + 3z^* = 12 + 8\text{i}$, where z^* is the complex conjugate of z. (4 marks)

3 Find the general solution in radians of the equation

$\tan\left(2x - \dfrac{\pi}{4}\right) = \sqrt{3}$. (6 marks)

4 A transformation is given by $\begin{bmatrix} x' \\ y' \end{bmatrix} = \mathbf{M}\begin{bmatrix} x \\ y \end{bmatrix}$.

Describe the geometrical transformation in the cases where:

(a) $\mathbf{M} = \begin{bmatrix} -1 & 0 \\ 0 & 1 \end{bmatrix}$, **(b)** $\mathbf{M} = \begin{bmatrix} 0 & -1 \\ 1 & 0 \end{bmatrix}$,

(c) $\mathbf{M} = \begin{bmatrix} -1 & 0 \\ 0 & 1 \end{bmatrix}\begin{bmatrix} 0 & -1 \\ 1 & 0 \end{bmatrix}$. (8 marks)

5 A curve satisfies the differential equation $\frac{dy}{dx} = \frac{1}{\sqrt{x^2 + 16}}$.
Starting at the point (3, 2.4) on the curve, use a step-by-step method with a step length of 0.25 to estimate the value of y at $x = 3.75$, giving your answer to two decimal places. (5 marks)

6 (a) Find the value of $\sum_{r=100}^{200} r^3$ (4 marks)

(b) Prove that $\sum_{r=1}^{n} (r^2 + 4r) = \frac{n}{6}(n + 1)(2n + 13)$. (4 marks)

7 A curve has equation $y = \frac{3x + 5}{x - 1}$.

(a) State the coordinates of the points where the curve crosses the coordinate axes. (2 marks)

(b) Write down the equations of its asymptotes. (2 marks)

(c) (i) Sketch the graph of $y = \frac{3x + 5}{x - 1}$, (2 marks)

(ii) Hence, or otherwise, solve the inequality
$\frac{3x + 5}{x - 1} > 1$. (3 marks)

8 A curve has equation $y = \frac{x^2 - 2x - 5}{x^2 + 2x + 2}$

(a) Write down the equation of the curve's asymptote. (1 mark)

(b) Prove that the curve only exists for $-3 \leqslant y \leqslant 2$. (6 marks)

(c) Hence determine the coordinates of the stationary points of the curve. (4 marks)

9 The variables x and y are known to satisfy an equation of the form $y = a + b\sqrt{x}$, where a and b are constants.

Corresponding approximate values of x and y (each rounded to one decimal place) were obtained experimentally and are given in the following table.

x	2.3	13.1	23.8	34.2	44.7	55.4	62.8
y	9.3	14.2	16.5	19.3	21.3	23.0	24.1

One of the values of y is known to have been recorded inaccurately.

(a) Draw a graph of y against $\sqrt{x}$. (4 marks)

(b) Identify the wrongly recorded value of y and estimate the correct value. (2 marks)

(c) Estimate the values of a and b, giving both answers to one decimal place. (4 marks)

Answers

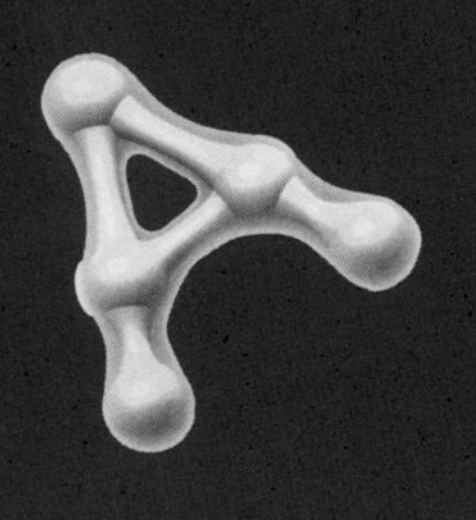

1 Roots of quadratic equations

EXERCISE 1A

1 **(a)** sum $= -4$, product $= -9$; **(b)** sum $= \frac{3}{2}$, product $= -\frac{5}{2}$;
(c) sum $= -5$, product $= -\frac{3}{2}$; **(d)** sum $= \frac{2}{3}$, product $= -\frac{1}{3}$;
(e) sum $= -\frac{12}{7}$, product $= -\frac{6}{7}$; **(f)** sum $= 3$, product $= -6$;
(g) sum $= -2$, product $= -2$; **(h)** sum $= a$, product $= -2a^2$;
(i) sum $= \dfrac{1-2a}{a}$, product $= 8$; **(j)** sum $= -2$, product $= -23$.

2 **(a)** $x^2 - 5x + 8 = 0$; **(b)** $x^2 + 3x + 5 = 0$;
(c) $x^2 - 4x - 7 = 0$; **(d)** $x^2 + 9x - 4 = 0$;
(e) $20x^2 - 5x + 8 = 0$; **(f)** $3x^2 + 2x + 12 = 0$;
(g) $5x^2 - 3x = 0$; **(h)** $x^2 - kx + 3k^2 = 0$;
(i) $x^2 - (k+2)x + 6 - k^2 = 0$; **(j)** $x^2 + (2 - a^2)x + (a+7)^2 = 0$.

EXERCISE 1B

1 **(a)** $\dfrac{2(\alpha+\beta)}{\alpha\beta}$; **(b)** $\dfrac{\alpha+\beta}{(\alpha\beta)^2}$;
(c) $\dfrac{(\alpha+\beta)^2 - 2\alpha\beta}{3\alpha\beta}$; **(d)** $\alpha\beta(\alpha+\beta)$;
(e) $4\alpha\beta - 2(\alpha+\beta) + 1$; **(f)** $\dfrac{(\alpha+\beta)^2 - 2\alpha\beta + 5(\alpha+\beta)}{\alpha\beta}$.

2 **(a)** -81; **(b)** -1.

3 **(a)** -4; **(b)** -56.

4 **(a)** $\frac{53}{4}$; **(b)** $-192\frac{1}{2}$.

5 **(a)** $\alpha + \beta = 5$, $\alpha\beta = 3$; **(b)** **(i)** -3 **(ii)** $8\frac{8}{9}$.

6 **(a)** $\frac{4}{3}$; **(b)** $\frac{10}{3}$; **(c)** 10; **(d)** 12; **(e)** 4; **(f)** 8.

7 **(a)** **(i)** -4, **(ii)** 1; **(b)** **(i)** 54, **(ii)** -2.

EXERCISE 1C

1 $x^2 - 44x + 16 = 0$. **2** $5x^2 - 6x - 4 = 0$.

3 $x^2 - 13x + 27 = 0$. **4** $8x^2 - 30x - 27 = 0$.

5 $27x^2 - 162x + 1 = 0$. **6** $x^2 + 10x + 1 = 0$.

7 **(a)** $\frac{16}{3}$; **(b)** $9x^2 - 66x + 61 = 0$.

8 **(a)** $-\frac{217}{8}$; **(b)** $18x^2 + 217x + 12 = 0$.

9 $16x^2 + 36x + 25 = 0$. **10** $x^2 - 24x - 106 = 0$.

EXERCISE 1D

1 $k = 20$. **2** $k = \pm 8$. **3** $7x^2 - 9x - 9 = 0$.

4 $2x^2 + 13x + 17 = 0$. **5** $x^2 - 16x + 40 = 0$.

6 **(a)** **(i)** 39, **(ii)** -230; **(b)** $7x^2 - 230x - 49 = 0$.

7 **(a)** **(i)** $\frac{22}{9}$, **(ii)** $-\frac{22}{27}$; **(b)** $81x^2 + 66x + 1 = 0$.

8 **(a)** **(i)** $\alpha + \beta = -2$, $\alpha\beta = 3$, **(iii)** $\frac{10}{27}$;
(b) $27x^2 - 10x + 1 = 0$.

9 **(a)** **(ii)** 18; **(b)** **(ii)** 47; **(c)** $x^2 - 15x - 45 = 0$.

10 **(a)** $\alpha + \beta = -3$, $\alpha\beta = -2$;
(b) **(i)** $\frac{13}{4}$, **(ii)** $-\frac{17}{4}$; **(c)** $4x^2 + 51x - 17 = 0$.

2 Complex numbers

EXERCISE 2A

1 **(a)** i; **(b)** -1; **(c)** i; **(d)** $-$i;
(e) i; **(f)** i; **(g)** -1.

2 **(a)** -8i; **(b)** 81; **(c)** -49;
(d) -4; **(e)** 27i; **(f)** -32i.

3 **(a)** -3i, 3i; **(b)** -10i, 10i; **(c)** -7i, 7i; **(d)** $-$i, i;
(e) -11i, 11i; **(f)** -8i, 8i; **(g)** $-n$i, ni.

4 **(a)** $-\sqrt{5}$i, $\sqrt{5}$i; **(b)** $-\sqrt{3}$i, $\sqrt{3}$i; **(c)** $-2\sqrt{2}$i, $2\sqrt{2}$i
(d) $-2\sqrt{5}$i, $2\sqrt{5}$i; **(e)** $-3\sqrt{2}$i, $3\sqrt{2}$i; **(f)** $-4\sqrt{3}$i, $4\sqrt{3}$i.

5 **(b)** Not real.

6 **(a)** $x^2 - 7x + 12 = 0$, $x = 3, 4$;
(b) $4x^2 - 19x + 24 = 0$, no real solutions;
(c) $x^2 - 17x + 60 = 0$, $x = 5, 12$.

EXERCISE 2B

1 **(a)** $3 + $i; **(b)** $2 - 6$i; **(c)** $-3 + 8$i; **(d)** $-7 - 5$i;
(e) $3 - \sqrt{2}$i; **(f)** $4 + \sqrt{3}$i; **(g)** $-1 + \frac{1}{3}$i.

2 **(a)** $8 - $i; **(b)** $-2 + 3$i; **(c)** $13 + 9$i;
(d) $5 - 39$i; **(e)** 27; **(f)** -12i.

3 **(a)** $30 - $i; **(b)** $27 + 11$i; **(c)** $65 + 17$i;
(d) $81 - 33$i; **(e)** $39 - 80$i; **(f)** $-7 - 24$i.

4 **(a)** $8 - 6$i; **(b)** $-32 + 24$i; **(c)** $-55 + 48$i;
(d) $24 - 70$i; **(e)** $7 + 6\sqrt{2}$i; **(f)** $13 - 8\sqrt{3}$i.

5 **(a)** 1; **(b)** $18 - 29$i; **(c)** $9 + 19$i.

6 **(a)** **(i)** 4, **(ii)** 6i, **(iii)** 13;
(b) **(i)** -8, **(ii)** 4i, **(iii)** 20;
(c) **(i)** -10, **(ii)** -6i, **(iii)** 34;
(d) **(i)** 12, **(ii)** -10i, **(iii)** 61;
(e) **(i)** $2x$, **(ii)** $2y$i, **(iii)** $x^2 + y^2$.

7 $p = -14$. **8** $q = -14$.

9 **(a)** $5 - $i; **(b)** 26.

10 **(a)** $3 + 4$i; **(b)** **(i)** $2 + 11$i, **(ii)** $-7 + 24$i.

EXERCISE 2C

1 $1 \pm 2i$.
2 $-2 \pm 3i$.
3 $1 \pm 3i$.
4 $3 \pm 4i$.
5 $4 \pm 2i$.
6 $-2 \pm i$.
7 $6 \pm 2i$.
8 $-1 \pm 7i$.
9 $-4 \pm i$.
10 $5 \pm 3i$.
11 $\frac{1}{2} \pm \frac{3}{2}i$.
12 $-\frac{1}{3} \pm i$.
13 $1 \pm \frac{1}{2}i$.
14 $\frac{3}{5} \pm \frac{4}{5}i$.
15 $-\frac{5}{13} \pm \frac{12}{13}i$.
16 $1 \pm \sqrt{3}i$.
17 $-2 \pm \sqrt{5}i$.
18 $3 \pm \sqrt{7}i$.
19 $-4 \pm \sqrt{3}i$.
20 $\frac{1}{2} \pm \frac{\sqrt{3}}{2}i$.

EXERCISE 2D

1 $a = 5, b = 2$.
2 $p = 5, q = 7$.
3 $t = 4, u = -24$.
4 $z = 4 - i$.
5 $z = 3 + 4i$.
6 $z = 2 + i$.
7 $3 + iy$, where y is real.
8 **(a)** $3 + 4i, -3 - 4i$; **(b)** $-6 + i, 6 - i$; **(c)** $5 + 2i, -5 - 2i$.
9 **(a)** $p = -8, q = 9$; **(b)** $1 + 2i$, coefficients are not real.
10 **(a)** $1 + 2i, -1 - 2i$; **(b)** $i, -2 - 3i$.

3 Inequalities

EXERCISE 3A

1 $x < 2, x > \frac{5}{2}$.
2 $4 < x < 7$.
3 $x < 2, x \geqslant \frac{11}{2}$.
4 $\frac{13}{2} < x < 7$.
5 $-4 < x < -\frac{7}{3}$.
6 $6 < x < \frac{29}{4}$.
7 $-4 < x < \frac{13}{5}$.
8 $0 \leqslant x < 3, x \geqslant 4$.
9 $\frac{5}{2} < x \leqslant 3, x \geqslant 5$.
10 $-8 < x < -1$ and $x > 4$.

EXERCISE 3B

1 $x < -6, x > \frac{5}{2}$.
2 $-\frac{13}{2} < x < 3$.
3 $-2 < x \leqslant 1, x \geqslant 2$.
4 $-2 < x < 2, x > 3$.
5 $-8 < x < -2, x > 1$.
6 $-4 < x < -2, -1 < x < 2$.
7 $x < -3, -2 < x < 0, x > 6$.
8 $0 \leqslant x < 3, x \leqslant -2$.
9 $-4 < x \leqslant -\frac{11}{9}, x > 1$.
10 $\frac{11}{5} < x < 3, x > 5$.

EXERCISE 3C

1 $-2 < x \leqslant 5$.
2 $x < -3, -1 < x \leqslant 0$.
3 $x \leqslant -\frac{9}{2}, x > 1$.
4 $0 < x < 4$.
5 $x < -6, -5 < x < 1$.
6 $\frac{9}{4} \leqslant x \leqslant 4, x \neq 3$.
7 $x < -\frac{2}{3}, 0 < x < 2, x > 3$.
8 $x < -5, -2 < x < 19$.
9 $2 < x < 4, -3 < x < 1$.
10 $x < -6, -5 < x < -3$.
11 $0 < x < 1, x > 2$.
12 $x < 2, x \geqslant 3$.
13 **(b)** Step 1; **(c)** $2 < x \leqslant 3$

4 Matrices

EXERCISE 4A

1 **(a)** 2×4; **(b)** 2×2; **(c)** 1×3; **(d)** 3×2.

2 **(a)** $\begin{bmatrix} 0 \\ 15 \end{bmatrix}$; **(b)** $\begin{bmatrix} 5 \\ -7 \end{bmatrix}$; **(c)** $\begin{bmatrix} 9 & 15 \\ -8 & 3 \end{bmatrix}$;

(d) $\begin{bmatrix} 4 & 4 \\ 4 & 11 \end{bmatrix}$; **(e)** $\begin{bmatrix} 4 & 23 \\ 14 & 3 \end{bmatrix}$.

3 **(a)** $\begin{bmatrix} 2 \\ 10 \end{bmatrix}$; **(b)** $\begin{bmatrix} 2 \\ -6 \end{bmatrix}$; **(c)** $\begin{bmatrix} 5 & -7 \\ -8 & 4 \end{bmatrix}$;

(d) $\begin{bmatrix} -2 & 0 \\ 10 & -10 \end{bmatrix}$; **(e)** $\begin{bmatrix} 0 & 13 \\ -1 & 2 \end{bmatrix}$.

4 **(a)** $\begin{bmatrix} 3 & 6 \\ 9 & 13 \end{bmatrix}$; **(b)** $\begin{bmatrix} 3 & 6 \\ 9 & 13 \end{bmatrix}$; **(c)** $\begin{bmatrix} 4 & -2 \\ -8 & -13 \end{bmatrix}$;

(d) $\begin{bmatrix} 7 & 4 \\ 1 & 0 \end{bmatrix}$; **(e)** $\begin{bmatrix} 9 & -12 \\ 9 & 21 \end{bmatrix}$; **(f)** $\begin{bmatrix} 13 & 1 \\ 10 & 17 \end{bmatrix}$;

(g) $\begin{bmatrix} 1 & 7 \\ -8 & -17 \end{bmatrix}$; **(h)** $\begin{bmatrix} -7 & 11 \\ 8 & 9 \end{bmatrix}$.

EXERCISE 4B

1 **(a)** $\begin{bmatrix} 32 \\ 19 \end{bmatrix}$; **(b)** $\begin{bmatrix} 45 \\ 90 \end{bmatrix}$; **(c)** $\begin{bmatrix} 57 \\ -8 \end{bmatrix}$;

(d) $\begin{bmatrix} -7 \\ 17 \end{bmatrix}$; **(e)** $\begin{bmatrix} 36 \\ 8 \end{bmatrix}$; **(f)** $\begin{bmatrix} -108 \\ 24 \end{bmatrix}$.

2 **(a)** $\begin{bmatrix} 42 & 18 \\ 81 & 15 \end{bmatrix}$; **(b)** $\begin{bmatrix} 3 & 54 \\ -2 & 18 \end{bmatrix}$; **(c)** $\begin{bmatrix} 18 & -26 \\ -20 & 46 \end{bmatrix}$;

(d) $\begin{bmatrix} -64 & 32 \\ -25 & 15 \end{bmatrix}$.

3 **(a)** $\begin{bmatrix} 7 & -1 \\ 2 & 14 \end{bmatrix}$; **(b)** $\begin{bmatrix} 23 & 11 \\ -22 & -54 \end{bmatrix}$.

4 **(a)** $x = -5, y = -10$; **(b)** $x = 64, y = -8$;

(c) $x = 4, y = -2$; **(d)** $x = 2, y = 1$;

(e) $x = 4, y = -6$; **(f)** $x = -2, y = 3$.

5 $c = 16$.

6 $\begin{bmatrix} 4 & 2 \\ -1 & 5 \end{bmatrix}$.

7 $\begin{bmatrix} 2 & -1 \\ 4 & 0 \end{bmatrix}$.

9 $\begin{bmatrix} \frac{3}{2} & -1 \\ -4 & 3 \end{bmatrix}$.

10 $\begin{bmatrix} -\frac{3}{10} & \frac{2}{5} \\ \frac{1}{10} & \frac{1}{5} \end{bmatrix}$.

5 Trigonometry

EXERCISE 5A

1 **(a)** $-\frac{\sqrt{3}}{2}$; **(b)** $-\sqrt{3}$; **(c)** $-\frac{1}{\sqrt{2}}$; **(d)** $\frac{1}{2}$;
(e) $-\frac{1}{\sqrt{3}}$; **(f)** $\frac{-\sqrt{3}}{2}$; **(g)** $-\frac{1}{2}+\frac{1}{\sqrt{3}}$; **(h)** $-\frac{1}{2}+\frac{1}{\sqrt{2}}$.

2 **(a)** $\frac{\sqrt{3}}{2}$; **(b)** -1 **(c)** $-\frac{\sqrt{3}}{2}$; **(d)** $-\frac{1}{2}$;
(e) $\sqrt{3}$; **(f)** $-\frac{1}{\sqrt{2}}$; **(g)** $-\frac{1}{2}$; **(h)** $-\sqrt{3}$;
(i) $-\frac{1}{2}$; **(i)** 0; **(k)** $-\frac{1}{\sqrt{3}}$; **(l)** -1.

4 $\frac{3\sqrt{3}}{2}$. **5** $\frac{7}{6}$. **6** $\frac{3+\sqrt{3}}{2}$.

EXERCISE 5B

Correct answers to general solutions may be given in a different form to those below; changing n to $n + 1$ or n to $n - 1$ will usually allow you to compare correct alternatives, for example in **1(c)**, a correct alternative to $(180n - 60)°$ is $(180n + 120)°$.

1 **(a)** $x = (360n + 45)°, (360n + 135)°$; **(b)** $x = (360n \pm 90)°$;
(c) $x = (180n - 30)°, (180n - 60)°$; **(d)** $x = (90n \pm 15)°$;
(e) $x = (180n + 18.4)°, (180n + 71.6)°$; **(f)** $x = (120n \pm 44.8)°$;
(g) $x = (360n + 10)°, (360n + 90)°$;
(h) $x = (15 - 180n)°, (35 - 180n)°$.

2 **(a)** $x = 2n\pi \pm \frac{\pi}{4}$;
(b) $x = 2n\pi, 2n\pi + \pi$ (Note: these can be combined to give $x = n\pi$);
(c) $x = \frac{n\pi}{2} \pm \frac{\pi}{24}$;
(d) $x = n\pi - \frac{\pi}{12}, n\pi + \frac{5\pi}{12}$;
(e) $x = n\pi \pm 0.4636^c$;
(f) $x = \frac{2n\pi}{3} - 0.5918^c, x = \frac{2n\pi}{3} + 0.9723^c$;
(g) $x = 2n\pi + \frac{\pi}{6}, x = 2n\pi + \frac{5\pi}{6}$;
(h) $x = -n\pi - \frac{\pi}{8}$ $\left(\text{This can be written as } x = n\pi - \frac{\pi}{8}\right)$.

3 $x = 2n\pi \pm \frac{\pi}{2}, 2n\pi + \frac{\pi}{6}, 2n\pi + \frac{5\pi}{6}$.

4 $x = (360n \pm 60)°, (360n \pm 90)°$.

5 $x = 2n\pi \pm \frac{\pi}{4}, 2n\pi \pm \frac{3\pi}{4}$.

6 $x = (180n \pm 30)°, (180n \pm 60)°$.

EXERCISE 5C

1 **(a)** $x = (180n + 20)°$; **(b)** $x = (180n - 45)°$;
(c) $x = (90n + 35)°$; **(d)** $x = (45n - 10)°$;
(e) $x = (180n + 25)°$; **(f)** $x = (-90n - 9.1)°$.

2 **(a)** $x = n\pi + \frac{\pi}{4}$; **(b)** $x = n\pi - \frac{\pi}{3}$;
(c) $x = \frac{n\pi}{4} - \frac{\pi}{16}$; **(d)** $x = \frac{n\pi}{2} + \frac{\pi}{12}$;
(e) $x = \frac{n\pi}{2} + \frac{5\pi}{12}$; **(f)** $x = \frac{7\pi}{24} - \frac{n\pi}{2}$.

3 $x = n\pi \pm \frac{\pi}{6}$.

4 $x = (180n - 45)°, (180n + 71.57)°$.

5 $x = n\pi$.

6 $x = (60n + 14)°$.

MIXED EXERCISE

1 **(a)** $\frac{\sqrt{3}}{3}$; **(b)** $x = (180n + 30)°$.

2 **(b)** $x = 2n\pi \pm \frac{\pi}{3}, 2n\pi \pm \pi$.

3 **(a)** $\frac{1}{2}, \frac{\sqrt{3}}{2}, \frac{1}{\sqrt{3}}$; **(b)** $x = n\pi \pm \frac{\pi}{6}$.

4 **(b)** $\sin x = -\frac{1}{2}, \sin x = 0; x = n\pi, 2n\pi - \frac{\pi}{6}, 2n\pi + \frac{7\pi}{6}$;
(c) $\theta = \frac{n\pi}{3}, \frac{2n\pi}{3} - \frac{\pi}{18}, \frac{2n\pi}{3} + \frac{7\pi}{18}$.

5 $\theta = (180n \pm 45)°$.

6 $x = (180n + 45)°, (360n \pm 60)°, (360n \pm 120)°$.

7 $\theta = 2n\pi \pm \frac{2\pi}{3}$.

6 Matrix transformations

EXERCISE 6A

1 **(a)** $A'(2, 4), B'(11, 19), C'(-6, -8)$; **(b)** $D(-2, 4)$.

2 **(a)** $P'(3, -6), Q'(-4, 9), R'(0, 6)$; **(b)** $S(4, 5)$.

3 $\begin{bmatrix} -1 & 2 \\ 3 & -1 \end{bmatrix}$.

4 **(a)** $\begin{bmatrix} 4 & 0 \\ 0 & 4 \end{bmatrix}$; **(b)** $\begin{bmatrix} 3 & 0 \\ 0 & 1 \end{bmatrix}$; **(c)** $\begin{bmatrix} 1 & 0 \\ 0 & 6 \end{bmatrix}$;
(d) $\begin{bmatrix} 8 & 0 \\ 0 & 8 \end{bmatrix}$; **(e)** $\begin{bmatrix} -2 & 0 \\ 0 & -2 \end{bmatrix}$; **(f)** $\begin{bmatrix} 4 & 0 \\ 0 & 2 \end{bmatrix}$.

5 **(a)** stretch in the x-direction, scale factor 3;
(b) stretch in the y-direction, scale factor 7;
(c) (two-way) stretch, scale factor 6 in the x-direction and scale factor 7 in the y-direction;
(d) enlargement, centre O, scale factor 5;
(e) stretch in the y-direction, scale factor 9;
(f) enlargement, centre O, scale factor -1;

(g) stretch in the x-direction, scale factor -3;

(h) (two-way) stretch, scale factor -4 in the x-direction and scale factor 9 in the y-direction;

(i) enlargement, centre O, scale factor 10;

(j) stretch in the y-direction, scale factor -4.

6 **(a)** (9, 12); **(b)** (6, 4); **(c)** (3, 20); **(d)** $(-12, 4)$.

EXERCISE 6B

1 **(a)** rotation of 180° about O;

(b) anti-clockwise rotation of 135° about O;

(c) reflection in the line $y = x \tan 30°$ ($y = \sqrt{3}x$);

(d) reflection in the line $y = x \tan 22.5°$;

(e) reflection in the line $y = x \tan 45°$ ($y = x$);

(f) clockwise rotation of 40° about O;

(g) reflection in the line $y = x \tan 35°$;

(h) anti-clockwise rotation of 26.6° about O;

(i) reflection in the line $y = x \tan 13.3°$;

(j) reflection in the line $y = x \tan 48.2°$;

(k) clockwise rotation of 53.1° about O;

(l) reflection in the line $y = x \tan 126.9°$.

2 **(a)** $\begin{bmatrix} \frac{1}{2} & \frac{\sqrt{3}}{2} \\ \frac{\sqrt{3}}{2} & -\frac{1}{2} \end{bmatrix}$; **(b)** $\begin{bmatrix} \frac{1}{2} & \frac{\sqrt{3}}{2} \\ -\frac{\sqrt{3}}{2} & \frac{1}{2} \end{bmatrix}$; **(c)** $\begin{bmatrix} \frac{1}{\sqrt{2}} & \frac{1}{\sqrt{2}} \\ -\frac{1}{\sqrt{2}} & \frac{1}{\sqrt{2}} \end{bmatrix}$;

(d) $\begin{bmatrix} -\frac{1}{2} & \frac{\sqrt{3}}{2} \\ \frac{\sqrt{3}}{2} & \frac{1}{2} \end{bmatrix}$; **(e)** $\begin{bmatrix} -\frac{1}{\sqrt{2}} & -\frac{1}{\sqrt{2}} \\ \frac{1}{\sqrt{2}} & -\frac{1}{\sqrt{2}} \end{bmatrix}$; **(f)** $\begin{bmatrix} \frac{1}{3} & \frac{2\sqrt{2}}{3} \\ -\frac{2\sqrt{2}}{3} & \frac{1}{3} \end{bmatrix}$;

(g) $\begin{bmatrix} -\frac{15}{17} & \frac{8}{17} \\ \frac{8}{17} & \frac{15}{17} \end{bmatrix}$; **(h)** $\begin{bmatrix} -\frac{4}{5} & -\frac{3}{5} \\ -\frac{3}{5} & \frac{4}{5} \end{bmatrix}$.

3 **(a)** $P'(-2 - \sqrt{3}, -2\sqrt{3} + 1)$;

(b) reflection in the line $y = x \tan 120°$ ($y = -x\sqrt{3}$).

4 **(a)** $\begin{bmatrix} \frac{1}{\sqrt{2}} & -\frac{1}{\sqrt{2}} \\ \frac{1}{\sqrt{2}} & \frac{1}{\sqrt{2}} \end{bmatrix}$; **(b)** $(2\sqrt{2}, 0)$;

(c) anti-clockwise rotation of 45° about O.

EXERCISE 6C

1 $\begin{bmatrix} \sqrt{3} & -1 \\ \frac{1}{2} & \frac{\sqrt{3}}{2} \end{bmatrix}$. **2** $\begin{bmatrix} 0 & 5 \\ 5 & 0 \end{bmatrix}$.

3 $\begin{matrix} -1 & 0 \\ 0 & 1 \end{matrix}$ **(a)** $\begin{bmatrix} & \\ & \end{bmatrix}$; **(b)** stretch in the x-dire

4 **(a)** $\begin{bmatrix} \frac{1}{2} & -\frac{\sqrt{3}}{2} \\ \frac{\sqrt{3}}{2} & \frac{1}{2} \end{bmatrix}$; **(b)** anti-clockwise rotation of 120° about O.

5 Enlargement, centre O, scale factor -1.

7 $\begin{bmatrix} \frac{1+2\sqrt{3}}{2\sqrt{2}} & \frac{2\sqrt{3}-1}{2\sqrt{2}} \\ \frac{\sqrt{3}-2}{2\sqrt{2}} & \frac{-\sqrt{3}-2}{2\sqrt{2}} \end{bmatrix}$.

8 Stretch in the y-direction of scale factor -1;

9 Anti-clockwise rotation of 53.1° about O followed by an enlargement, centre O, scale factor 5.

MIXED EXERCISE

1 **(a)** A′(3, −4), B′(4, 3);
   (b) clockwise rotation of 53.1° about O.

2 reflection in line $y = \frac{1}{2}x$.

3 **(a)** $\cos 2\theta = \frac{2}{3}$;
   (b) reflection in the line $y = x \tan 48.2°$ followed by an enlargement, centre O, scale factor 3;
   (c) enlargement, centre O, scale factor 9.

4 (7, 6).

7 Linear laws

EXERCISE 7A

1 **(a)** $y = \frac{1}{2}x + 3$; **(b)** 74.

2 **(a)** $S = 9 - 0.04t$; **(b)** 225.

3 **(a)** $R = 62 - 0.05L$; **(b)** 57.8.

4 **(a)** $p = 4,\ q = 3$; **(b)** $p = -3,\ q = 8$; **(c)** $p = -0.15,\ q = 40$.

EXERCISE 7B

1 (Note that there are alternative correct answers.)
   (a) Plot y against x^3, gradient of line gives a, intercept on y-axis gives b;
   (b) Plot y against $\sqrt{x}$, gradient of line gives b, intercept on y-axis gives a;
   (c) Plot y^2 against x, gradient of line gives a, intercept on y^2-axis gives b;
   (d) Plot $\frac{1}{y}$ against $\frac{1}{\sqrt{x}}$, gradient of line gives b, intercept on $\frac{1}{y}$-axis gives a;
   (e) Plot $\frac{y}{x}$ against x^2, gradient of line gives a, intercept on $\frac{y}{x}$-axis gives b;
   (f) Plot $\frac{y}{x}$ against $\frac{y^2}{x}$, gradient of line gives b, intercept on $\frac{y}{x}$-axis gives a;
   (g) Plot $\frac{y}{x}$ against $\frac{1}{x^2}$, gradient of line gives a, intercept on $\frac{y}{x}$-axis gives b;
   (h) Plot xy against x^2, gradient of line gives a, intercept on xy-axis gives b;
   (i) Plot y against $\frac{x^2}{y}$ gradient of line gives $-b$, intercept on y-axis gives a.

2 **(a)** $y = \frac{3}{2}x^2 + 2x$.

3 **(b)** $a \approx 8,\ b \approx \frac{1}{2}$.

4 (b) (i) ≈ 3.72, **(ii)** ≈ 1.38

5 (a) $y = \frac{3.5}{x} - 0.5$.

6 $a \approx 1.7$, $b \approx 1.3$.

EXERCISE 7C

1 $y = 10x^{0.75}$. **2** $y = 0.01x^2$. **3** $a \approx 4.0$, $b \approx 1.2$.

4 (a) $\log_{10} V = \log_{10} a - n \log_{10} T$;
(c) (i) 0.87, **(ii)** $a \approx 3.1$, $n \approx 0.50$.

5 (a) $\log_{10} Q = \log_{10} a + b \log_{10} x$;
(c) (i) 3.6 (or 3.7), **(ii)** $a = 16$ (or 17), $b \approx 2.5$.

EXERCISE 7D

1 $y \approx 3.16x^{3.16}$. **2** $P \approx 10 \times 0.708^T$.

3 (a) $\log_{10} T = \log_{10} k + L \log_{10} a$;
(c) (i) 83, **(ii)** $a = 3.2$, $k \approx 4.7$.

4 $a \approx 7.5$, $b \approx 2.0$.

5 $h \approx 2.4$, $k \approx 2.5$.

6 $p \approx 24$, $q \approx 1.6$.

MIXED EXERCISE

1 (b) (i) $b \approx 3.2$, **(ii)** $a \approx 5.0$ (or 5.1), $b \approx 0.33$.

2 $a \approx -6.2$.

3 (a) $\frac{y}{x^2} = ax + b$; **(c)** $a \approx 4.0$, $b \approx 2.0$.

4 $a = 1.8$, $b = 2.5$.

5 (a) 331;
(b) (i) $\log_{10} N = \log_{10} a + t \log_{10} b$; **(ii)** $a = 251$, $b = 1.06$;
(c) e.g. growth limited by test tube.

6 (b) $k \approx 32\,000$, $n \approx -1.4$ (or -1.3).

8 Calculus

EXERCISE 8A

1 (a) $6 + h$; **(b)** 6.

2 (b) 6.

3 (b) $2 + h$; **(c)** 2.

4 (a) $1 - h$; **(b)** 1.

5 (a) $h - 3$; **(b)** -3.

6 (a) $3h - 19$; **(b)** -19.

7 (a) $8 - 2h$; **(b)** 8.

EXERCISE 8B

1 (a) $h^2 + 6h + 12$; **(b)** 12.

2 (b) 12.

3 (b) $h^3 + 8h^2 + 24h + 27$; **(c)** 27.

4 (a) $-h^2 + 2h + 15$; **(b)** 15.

5 (a) $h^3 - 5h^2 + 9h - 7$; **(b)** -7.

6 (a) $4h^4 + 20h^3 + 40h^2 + 40h + 20$; **(b)** 20.

7 (a) $2h^3 + 21h^2 + 78h + 123$; **(b)** 123.

8 (a) $3h^5 - 18h^4 + 45h^3 - 60h^2 + 45h - 18$; **(b)** -18.

EXERCISE 8C

1 (a) (i) $\frac{1}{2} - \frac{1}{a}$, **(ii)** $\frac{1}{3} - \frac{1}{3b^3}$, **(iii)** $2 - \frac{1}{c}$, **(iv)** $\frac{d^2}{2} - \frac{1}{2}$;

(b) (i) $\frac{1}{2}$, **(ii)** $\frac{1}{3}$, **(iii)** 2, **(iv)** does not exist.

2 (a) (i) $\frac{1}{a} + \frac{1}{3}$, **(ii)** $\frac{1}{4b^4} - \frac{1}{4}$, **(iii)** $\frac{1}{4} - \frac{c^4}{4}$, **(iv)** $\frac{1}{5d^5} + \frac{32}{5}$;

(b) (i) $\frac{1}{3}$, **(ii)** $\frac{1}{4}$, **(iii)** does not exist, **(iv)** $\frac{32}{5}$.

3 (a) (i) $6 - 2\sqrt{a}$, **(ii)** $\frac{1}{b} - \frac{1}{2}$, **(iii)** $6 - \frac{3c^{\frac{2}{3}}}{2}$, **(iv)** $\frac{32}{3} - \frac{4d^{\frac{3}{4}}}{3}$;

(b) (i) 6, **(ii)** does not exist, **(iii)** 6, **(iv)** $\frac{32}{3}$.

4 $\frac{1}{\sqrt{x}}$ is not defined at lower limit when $x = 0$, 18.

5 $\frac{1}{x^2}$ is not defined when $x = 0$ which is part of the interval of integration.

9 Series

EXERCISE 9A

1 $2 + 3 + 4 + 5 + 6 = 20$.

2 $7 + 9 + 11 + 13 + 15 = 55$.

3 $5 + 10 + 17 + 26 = 58$.

4 $-5 - 1 + 3 + 7 + 11 = 15$.

5 $8 + 27 + 0 + 0 + 0 + 64 + 125 = 224$.

6 $0 + 2 + 6 + 12 + 20 = 40$.

7 $-24 - 6 + 6 + 24 = 0$.

8 $0 + 0 + 0 + 6 + 24 = 30$.

9 $0 + 0 + 4 + 18 + 48 + 100 = 176$.

10 $-60 - 24 - 6 + 0 + 0 + 0 + 6 = -84$.

EXERCISE 9B

1 (a) 50; **(b)** 2870; **(c)** 672 400;
(d) 295 425; **(e)** 47 002 725; **(f)** 71 894 400.

2 (a) 2310; **(b)** 13 395 600; **(c)** 24 645;
(d) 1 611 000; **(e)** 10 300; **(f)** 115 220.

3 72 880.

4 2 731 516.

5 (a) 9140; **(b)** 186 735.

6 (a) 375 750; **(b)** 292 291 750.

7 14.

8 (a) (i) 25 502 500, **(ii)** 23 876 875; **(b)** 3775; **(c)** 0.

EXERCISE 9C

6 $n(n+1)(5n^2+5n+1)$.

7 **(a)** $n(n+1)(n^2+n+1)$; **(b)** $n(n+1)(2n+3)$;
(c) $n(n+1)(2n^2+2n-1)$; **(d)** $n(n+1)(2n^2-1)$;
(e) $n(n+1)(3n^2+3n+4)$; **(f)** $3n(n-5)$.

10 Numerical methods

EXERCISE 10A

5 root between 1 and 2.

6 -1 and -2.

EXERCISE 10B

1 **(b)** $-2.5<\alpha<-2.4$.

2 **(b)** $1.5<\alpha<1.6$.

3 **(b)** $1.70<\alpha<1.75$.

4 **(b)** $0.6375<\alpha<0.6500$.

5 Graphs cross once only.
(b) $0.5<\alpha<0.6$.

6 $5.750<\alpha<5.875$.

7 **(b)** $0.850<\alpha<0.875$.

8 **(b)** $-1.975<\alpha<-1.950$.

9 **(a)** $x=2.81$; **(b)** **(ii)** $2.2<\alpha<2.3$.

EXERCISE 10C

1 **(b)** $\frac{1}{3}$.

2 $\frac{440}{215}\approx 2.05$.

3 3.5.

4 -4.07.

5 1.38.

6 1.52.

EXERCISE 10D

1 **(a)** -2.95; **(b)** 1.18; **(c)** 2.02.

2 Student A will not be successful since $f'(-1)=0$. Student B obtains a next approximation of $-2.417\ldots$ and after several iterations -2.196 (to 3 d.p.).

3 $x_2=-0.531\,25$, $x_3=-0.530\,00$.

4 **(b)** $x_2=-2.773$.

5 **(b)** -0.39; **(c)** ~~$-0.380\,72$.~~ -0.38829

6 **(b)** $-1.75<\alpha<-1.50$; **(c)** -1.4 **(d)** -1.72.

EXERCISE 10E

1 2.12. **2** 4.962. **3** 4.336. **4** 1.721. **5** 1.007.

6 **(a)** 0.27; **(b)** 0.277;
(c) Error in **(a)** = 4%, error in **(b)** = 1%.

11 Asymptotes and rational functions

EXERCISE 11A

1 $x=7$.

2 $x=2$.

3 $x=-1$.

4 $x=1$ and $x=0$.

5 $x=3$ and $x=-4$.

6 None.

7 $x=1$

8 $x=6$ and $x=-1$

9 None.

EXERCISE 11B

1 **(a)** $(0, \frac{5}{2})$ and $(5, 0)$; **(b)** $x = -2$ and $y = -1$;

(c)

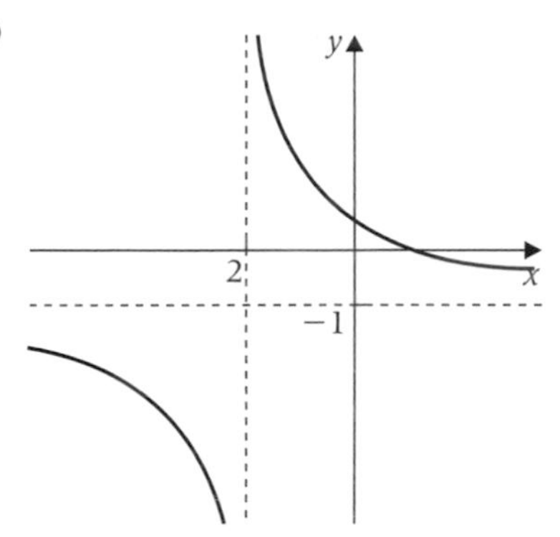

2 **(a)** $(0, -\frac{1}{2})$ and $(-1, 0)$ **(b)** $x = \frac{2}{3}$ and $y = \frac{1}{3}$

(c)

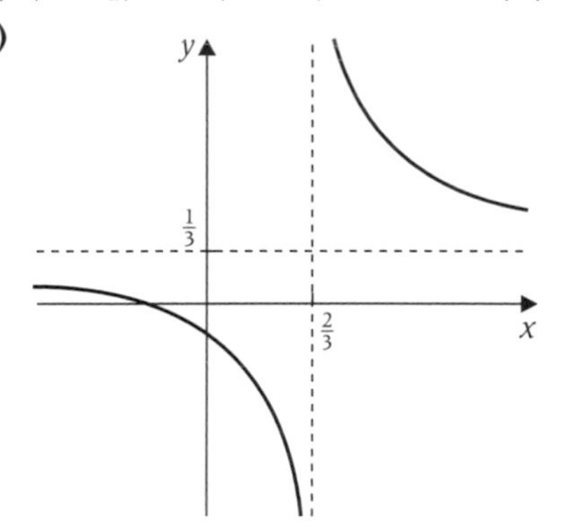

3 **(a)** **(i)** $x = 4$ and $y = \frac{1}{2}$, **(ii)** $(0, -\frac{3}{8})$ and $(-3, 0)$,

(iii)

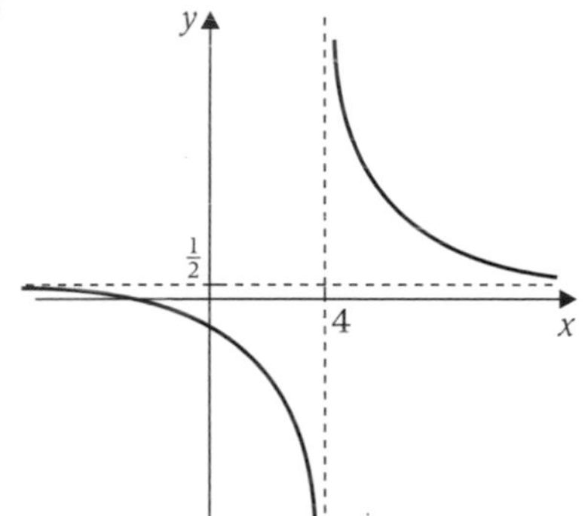

(b) **(i)** $x = -\frac{1}{2}$ and $y = 3$, **(ii)** $(0, -12)$ and $(2, 0)$,

(iii)

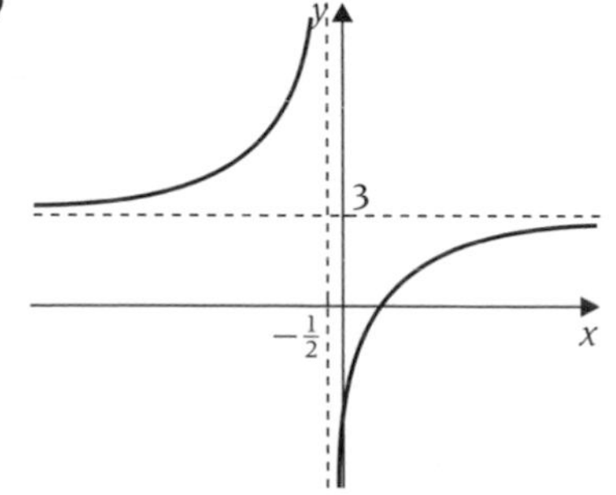

(c) (i) $x = 2$ and $y = 1$, **(ii)** $(0, -\frac{1}{2})$ and $(-1, 0)$,

(iii)

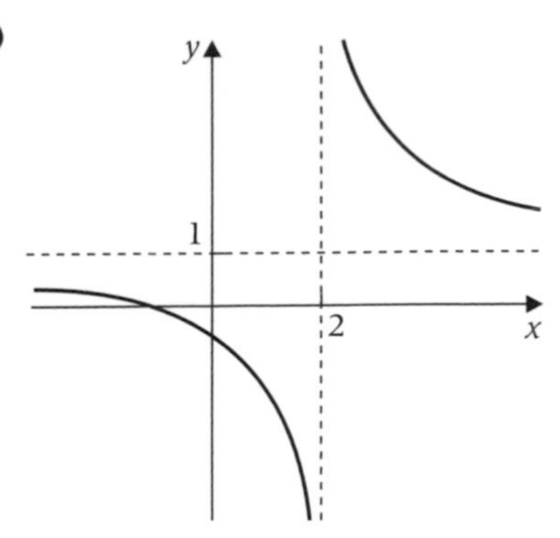

4 (a) (i) $x = -\frac{5}{2}$ and $y = \frac{1}{2}$, **(ii)** $(0, -\frac{3}{5})$ and $(3, 0)$,

(iii)

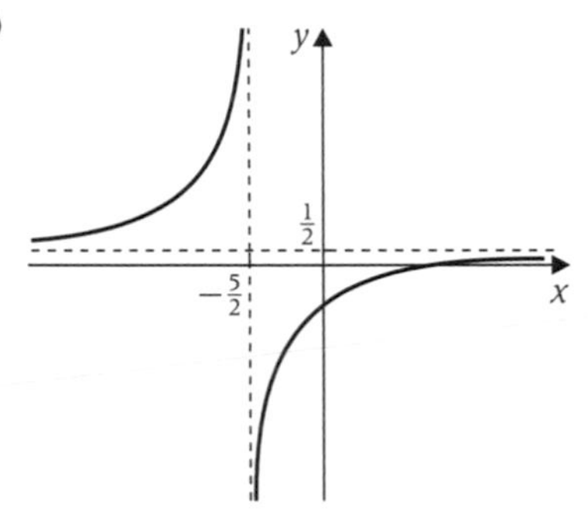

(b) (i) $x = \frac{1}{3}$ and $y = -1$, **(ii)** $(0, 4)$ and $(-\frac{4}{3}, 0)$,

(iii)

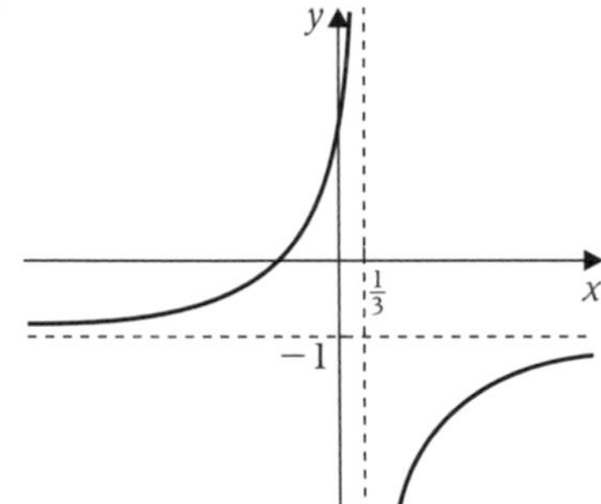

(c) (i) $x = 2$ and $y = 3$, **(ii)** $(0, -2)$ and $(-\frac{4}{3}, 0)$,

(iii)

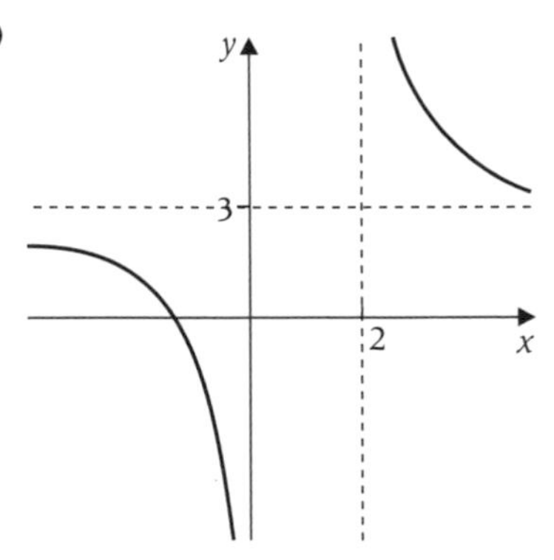

5 **(a)** $x = \frac{7}{4}$ and $y = \frac{5}{4}$,
$(0, -\frac{3}{7})$ and $(-\frac{3}{5}, 0)$

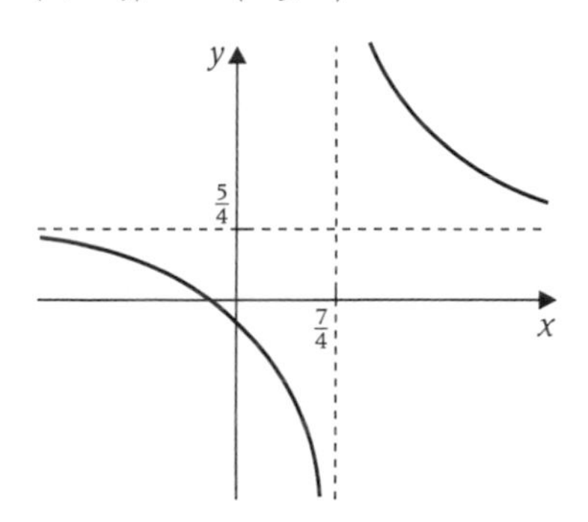

(b) $x = \frac{3}{5}$ and $y = \frac{8}{5}$
$(0, \frac{4}{3})$ and $(\frac{1}{2}, 0)$

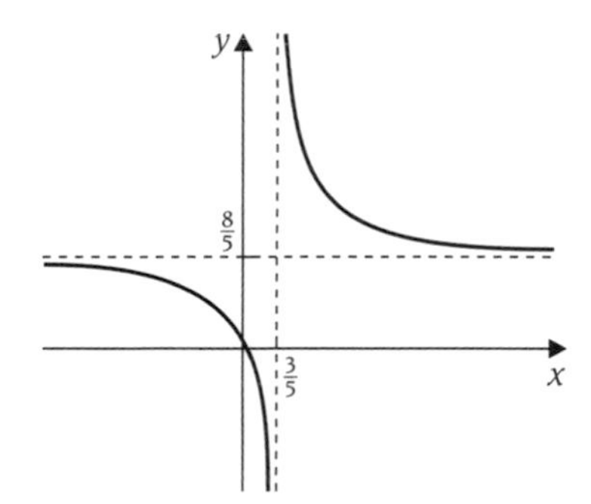

(c) $x = \frac{2}{3}$ and $y = \frac{1}{3}$
$(0, \frac{7}{2})$ and $(7, 0)$

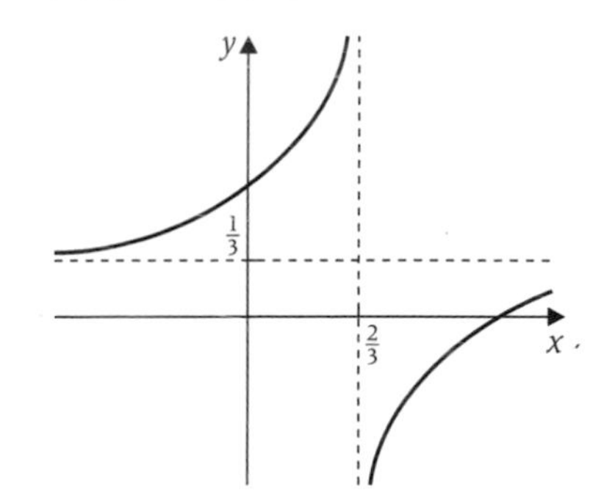

6 $a = 4$, $b = 1$.

EXERCISE 11C

1 $(1, 1)$ and $(-1\frac{1}{2}, -4)$.

2 $(-1, -2)$ and $(-\frac{2}{3}, -1)$.

3 $(2, 2)$ and $(-1, 5)$.

4 $(-1, -2)$, $(-\frac{5}{6}, -\frac{3}{2})$.

5 $(\frac{1}{2}, -1)$ and $(1, 1)$.

EXERCISE 11D

1 $x < -2$ and $x > -1$.

2 $\frac{2}{3} < x \leqslant 4$

3 $x < 3$ and $x > 4$.

4 $x < -\frac{1}{2}$ and $x > 2$.

5 $-\frac{5}{2} < x < -2$.

6 $-\frac{1}{2} < x < \frac{1}{3}$.

7 $\frac{7}{4} < x \leqslant \frac{17}{3}$.

8 $x < \frac{1}{3}$ and $x > \frac{3}{5}$.

9 **(a)** $x = \frac{5}{2}$ and $y = -2$

$(0, -\frac{3}{5})$ and $(\frac{3}{4}, 0)$

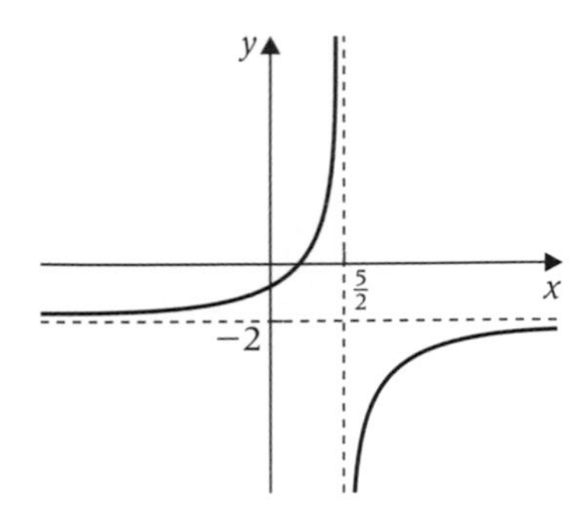

(b) $x < \frac{3}{4}$ and $x > \frac{5}{2}$.

10 **(a)** $x = 2$ and $y = 3$

$(0, -2)$ and $(-\frac{4}{3}, 0)$

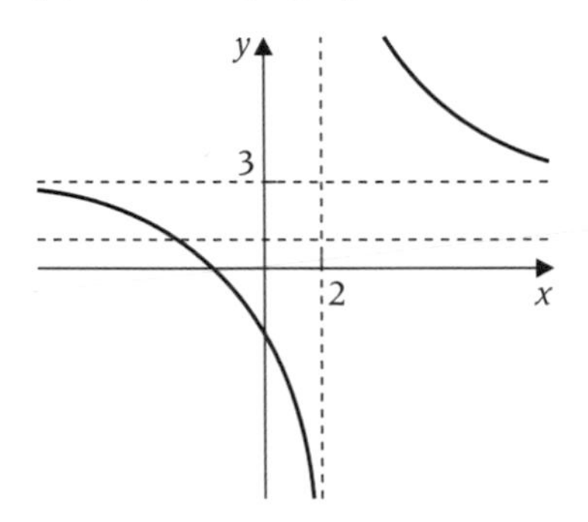

(b) $x < -3$ and $x > 2$.

11 **(a)** $x = 1$ and $y = 4$

$(0, 3)$ and $(\frac{3}{4}, 0)$

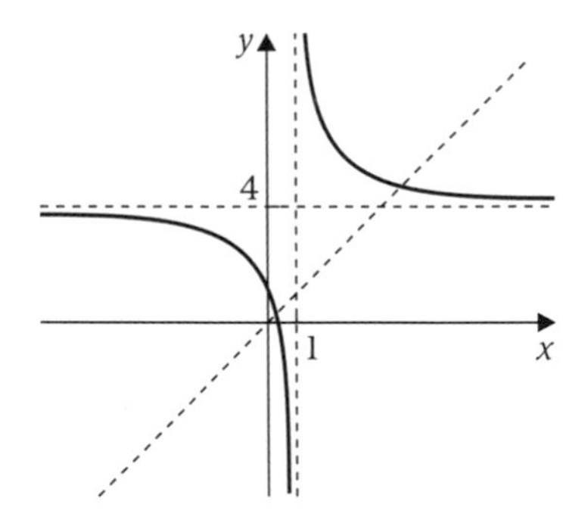

(b) $0 < x < 1$ and $x > 2$.

12 Further rational functions

EXERCISE 12A

1 $x = 3, x = 4, y = 0$

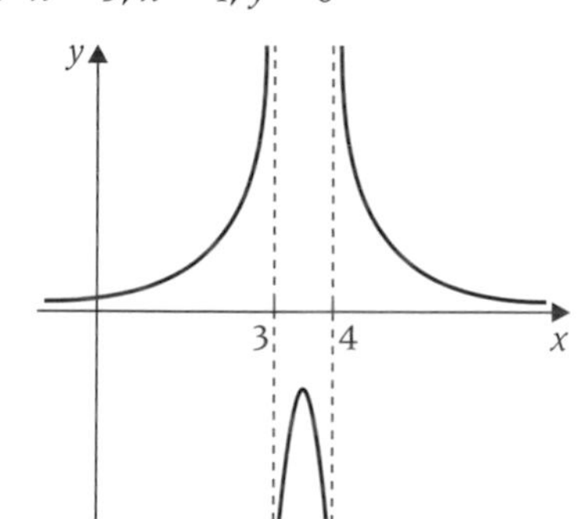

2 $x = 2, x = 3, y = 0$

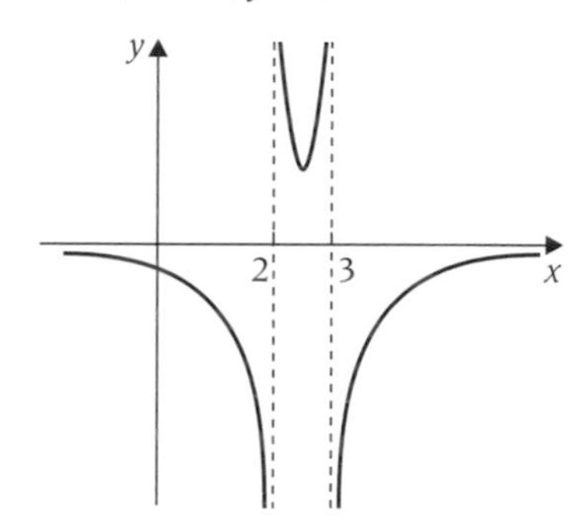

3 $x = 2, x = -2, y = 0$

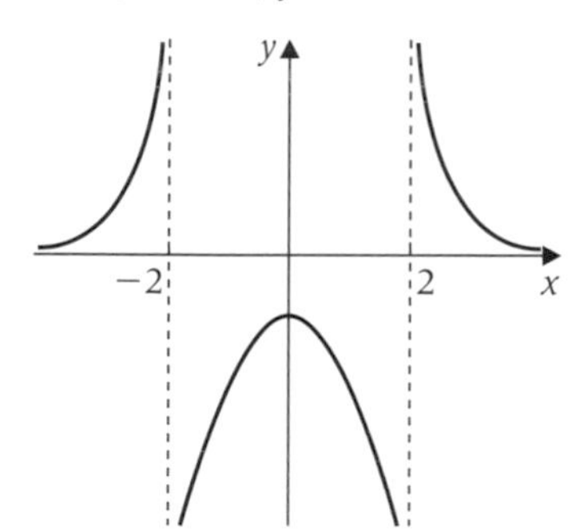

4 $x = 1, x = 3, y = 0$

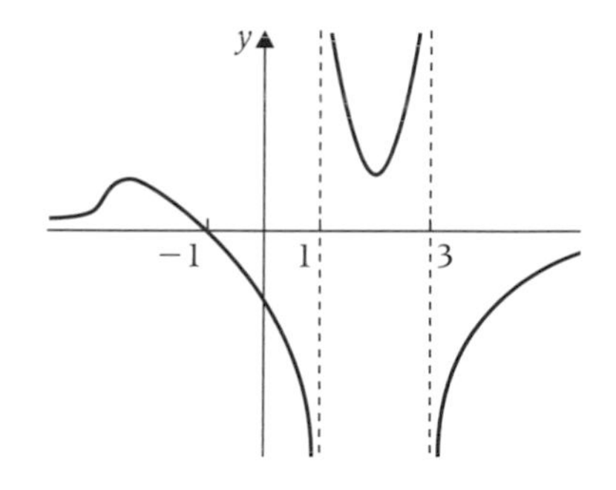

5 $x = 3, x = -2, y = 0$

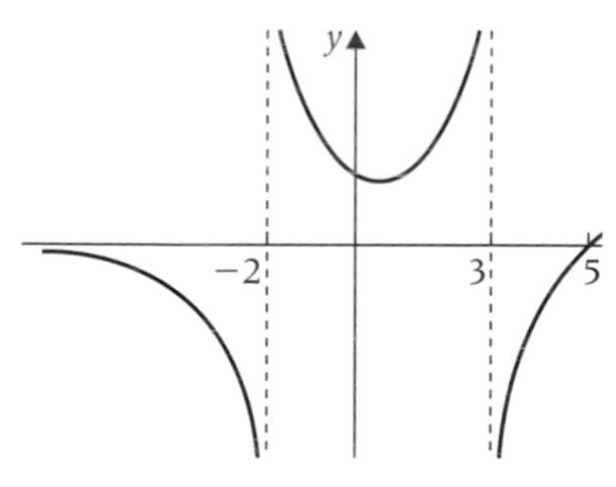

6 $x = 0, x = 4, y = 0$

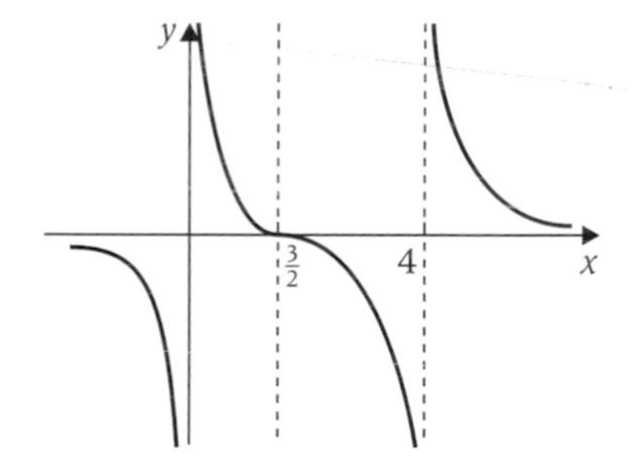

7 **(a)** $x = 5, x = -2, y = 0$; **(b)** $y = 0$.

8 **(a)** $(-1, \frac{1}{2}), (4, \frac{1}{2})$; **(b)** $(1, 1)$; **(c)** None.

EXERCISE 12B

1 **(a)** $x = 0, x = -\frac{4}{3}, y = \frac{1}{3}$; **(d)** $(-\frac{1}{2}, -1)$ and $(2, \frac{1}{4})$.

2 **(b)** $(-\frac{3}{2}, -3)$ and $(-4, 2)$.

3 $(0, -2)$ and $(-2, 0.4)$.

4 **(b)** $(1, -1)$ and $(-3, -\frac{1}{9})$; **(c)** $x = 0, x = 3, y = 0$.

5 **(b)** $(0, 0)$ and $(-2, 4)$; **(c)** $y = 1$, no vertical asymptotes.

6 **(a)** $x = \pm\sqrt{3}, y = 0, (-2, 0)$ and $(0, \frac{2}{3})$;

(b) $y \leqslant \frac{1}{6}$ and $y \geqslant \frac{1}{2}$;

(c) $(-3, \frac{1}{6})$ and $(-1, \frac{1}{2})$.

7 **(b)** $(\frac{5}{2}, \frac{2}{9})$ and $(\frac{1}{2}, 2)$;
(c) $x = -\frac{1}{2}, x = 1, y = 0$.

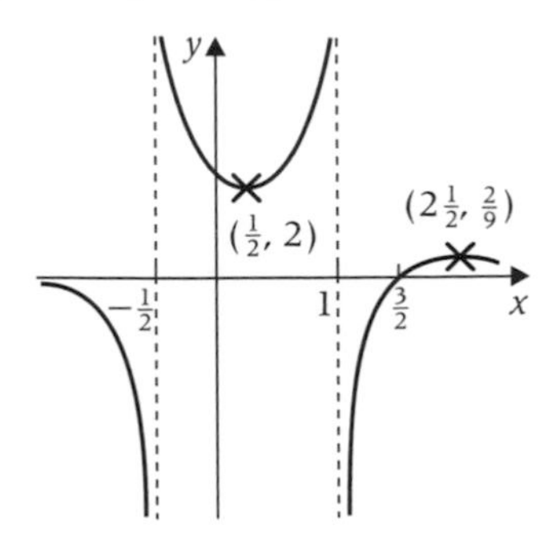

8 **(b)** $(-3, 2)$ and $(\frac{1}{5}, -\frac{18}{7})$;
(c) $y = 1$, no vertical asymptotes

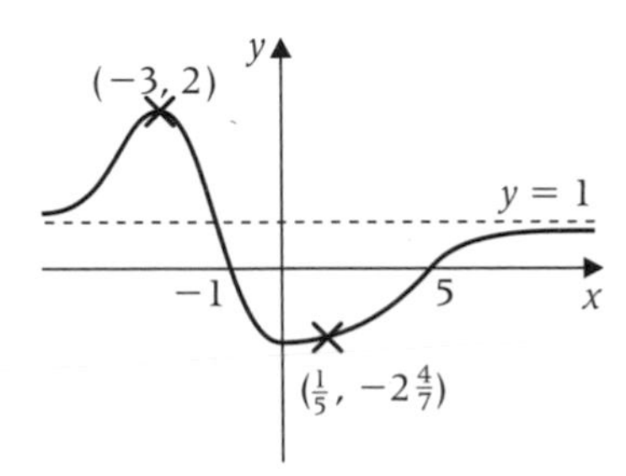

9 **(a)** $-5 \leqslant k \leqslant 1$; **(b)** $-5 \leqslant y \leqslant 1$;
(c) $(-3, 5)$ and $(3, 1)$

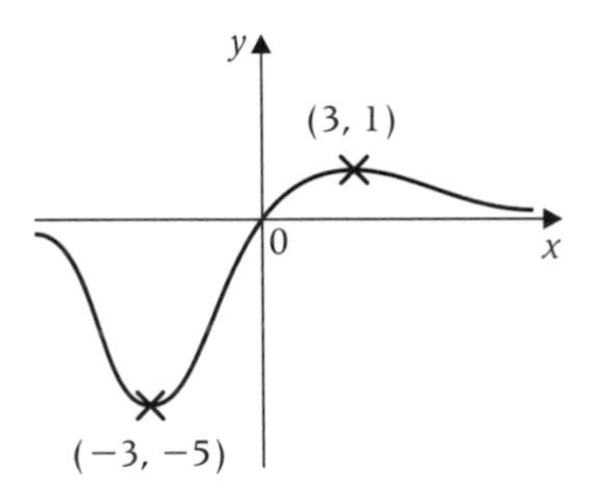

10 **(b)** minimum $(1, \frac{1}{3})$; maximum $(-1, 3)$;
(c) $y = 1, x^2 + x + 1 = 0$ has no real roots;
(d)

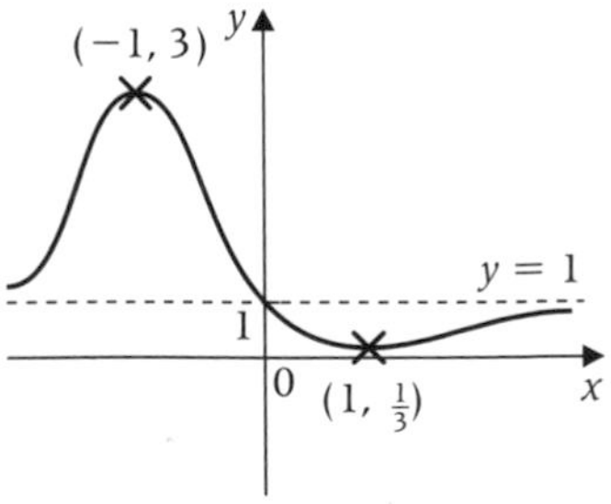

(−1, 3) y

y = 1

1

0 $(1, \frac{1}{3})$ x

13 Parabolas, ellipses and hyperbolas

EXERCISE 13A

1 $x = (y + 3)^2$

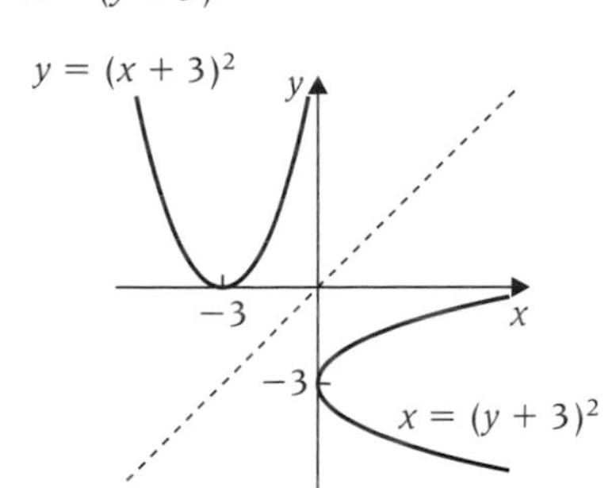

2 (a) C_2: $x = 4y^2 - 1$

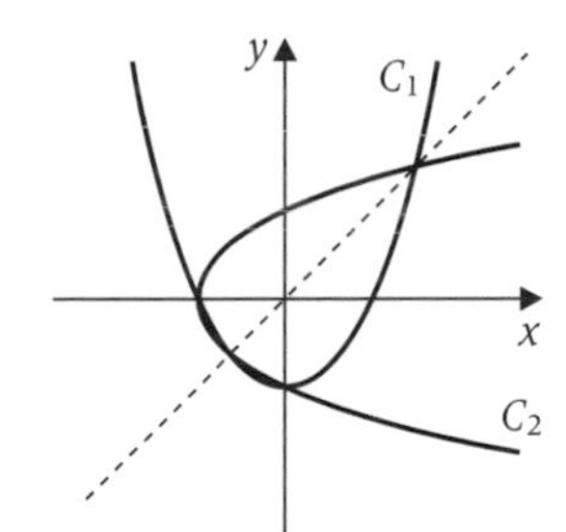

(b) C_3: $y = x^2 - 1$ **(c)** C_4: $y = 2x^2 - \frac{1}{2}$.

3 (a) $y^2 + 4x^2 = 16$

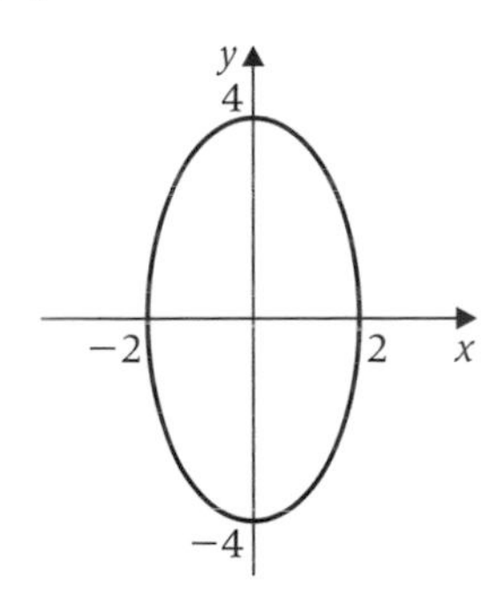

(b) $(0, 4)$, $(0, -4)$, $(2, 0)$ and $(-2, 0)$.

4 $x^2 + 4y^2 = 36$, $(-6, 0)$, $(6, 0)$, $(0, 3)$, $(0, -3)$

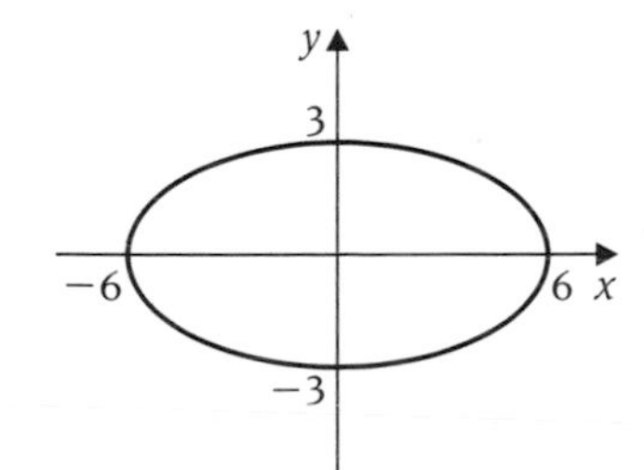

5 (a) Reflect in line $y = x$; then stretch by scale factor 2 in the y-direction;

(b) Stretch by scale factor 4 in the y-direction;

(c) Stretch by scale factor 2 in the x-direction.

6 **(a)** Stretch by scale factor 3 in the x-direction;

(b) Reflection in the line $y = x$;

(c) Stretch by scale factor 2 in the y-direction..

EXERCISE 13B

1 $y^2 - x^2 = 25$

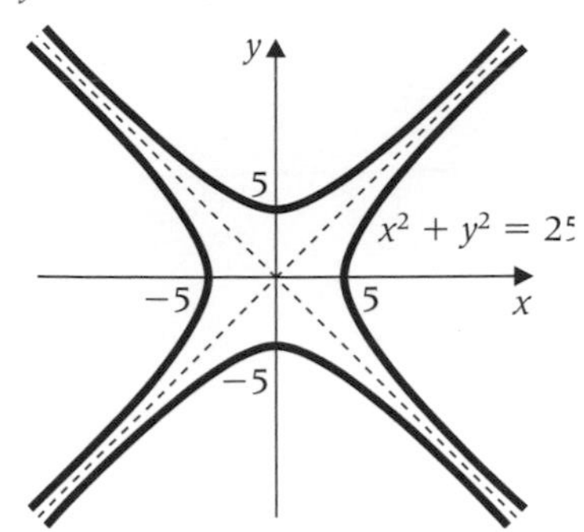

2 **(a)** H_2: $\dfrac{y^2}{4} - \dfrac{x^2}{9} = 1$

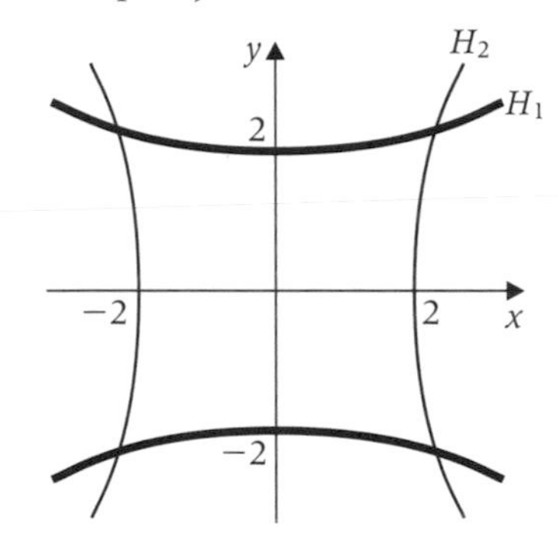

(b) H_3: $\dfrac{x^2}{4} - \dfrac{y^2}{36} = 1$; **(c)** H_4: $x^2 - \dfrac{y^2}{9} = 1$.

3 **(a)** $y^2 - 4x^2 = 16$

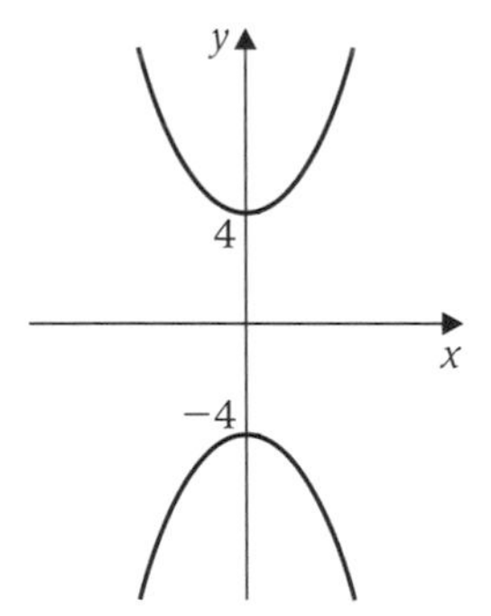

(b) $(0, 4)$, $(0, -4)$.

4 **(a)** Stretch by scale factor 3 in the x-direction;

(b) Reflection in $y = x$;

(c) Stretch in x-direction with scale factor $3x$ and stretch in y-direction with scale factor $\dfrac{1}{x}$ for any non-zero value of x;

(d) Stretch by scale factor $\dfrac{\sqrt{3}}{2}$ in the y-direction.

5 **(a)** $xy = 10$;
(b) **(i)** $x^2 - 4y^2 = 40$, **(ii)** No.

EXERCISE 13C

1 $\frac{(x+3)^2}{2} + \frac{(y-2)^2}{3} = 1$.

2 **(a)** $\frac{(x-3)^2}{4} - \frac{y^2}{5} = 1$; **(b)** $5(x+3)^2 + 7(y+2)^2 = 12$;
(c) $(x-2)^2 - (y+5)^2 = 3$ **(d)** $xy = 3$;
(e) $(y+1)^2 = x$; **(f)** $(x-3)^2 - y^2 = 7$.

3 Translation through vector $\begin{bmatrix} 5 \\ -4 \end{bmatrix}$.

4 $k = -7$, $\begin{bmatrix} 3 \\ 1 \end{bmatrix}$.

5 Translation with vector $\begin{bmatrix} -\frac{5}{3} \\ 2 \end{bmatrix}$.

6 $k = 2$, $\begin{bmatrix} 3 \\ -3 \end{bmatrix}$.

EXERCISE 13D

1 **(a)** $(\frac{17}{3}, \frac{19}{3})$ and $(1, -3)$; **(b)** $(-\sqrt{17}, -5)$, $(\sqrt{17}, -5)$.

3 **(a)** Intersect at $(1, 0)$ and $(-\frac{1}{2}, \frac{1}{2})$; **(b)** $-2 < k < 2$.

4 $m = \pm 2$.

5 **(a)** $(\frac{4}{9}, -\frac{2}{3})$ and $(1, 1)$; **(b)** $(-\frac{14}{13}, -\frac{68}{13})$ and $(2, 4)$;
(c) $(-1, -5)$ and $(-11, -35)$.

6 $p > 5, p < -1$.

7 **(a)** Does not intersect; **(b)** $k \leqslant 7$.

Exam style practice paper

1 **(a)** **(i)** $\alpha + \beta = 3$; $\alpha\beta = 4$, **(iii)** $-\frac{9}{8}$; **(b)** $8x^2 + 9x + 8 = 0$.

2 **(a)** $-2 \pm 3i$; **(b)** $3 - 4i$.

3 $\frac{7}{24}\pi + n\frac{\pi}{2}$.

4 **(a)** Reflection in the y-axis;
(b) Anticlockwise rotation of 90° about the origin;
(c) Reflection in the line $y = x$.

5 2.51.

6 **(a)** 379 507 500.

7 **(a)** $(-\frac{5}{3}, 0)$ and $(0, -5)$; **(b)** $x = 1, y = 3$;

(c) **(i)**

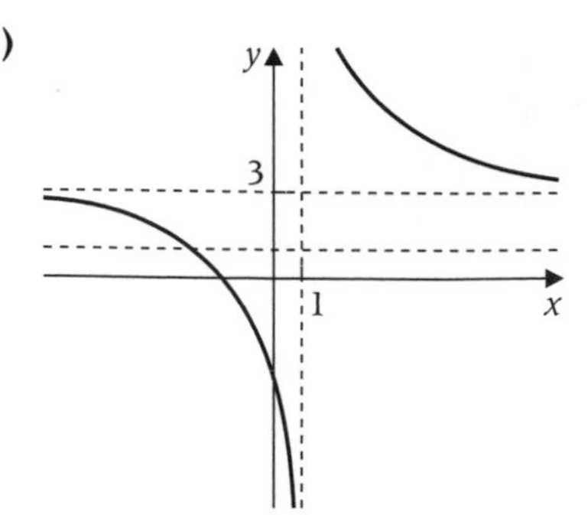

(ii) $x < -3$ and $x > 1$.

8 **(a)** $y = 1$; **(c)** $(-\frac{1}{2}, -3)$ and $(-3, 2)$.

9 **(a)**

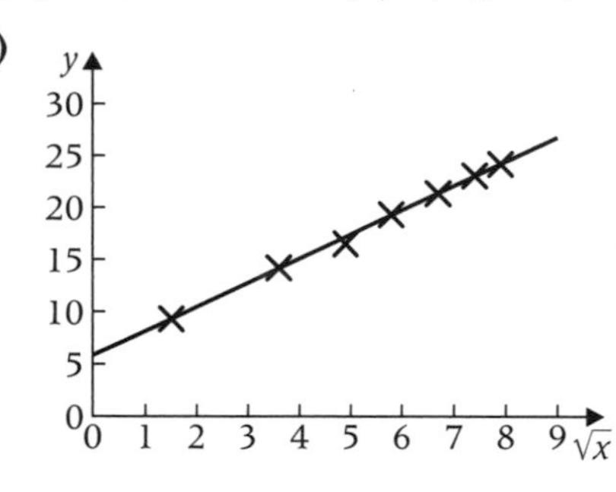

(b) 16.5 is wrong; 17.2; **(c)** $a = 5.8$, $b = 2.3$.

Index

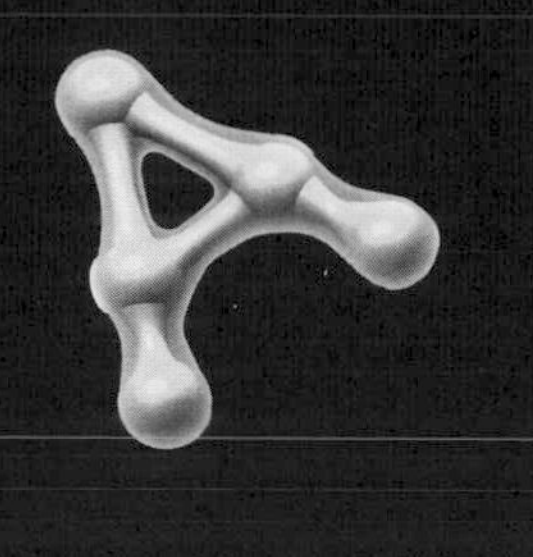